AMERICAN FLIGHT
JACKETS, AIRMEN
& AIRCRAFT

AMERICAN FLIGHT JACKETS, AIRMEN & AIRCRAFT

A History of U.S. Flyers' Jackets from World War I to Desert Storm

Jon A. Maguire & John P. Conway

Schiffer Military/Aviation History
Atglen, PA

Acknowledgements

Assembling the photographs and material for this work involved a tremendous amount of communication with historians, veterans, authors, and collectors. The end result is due to the generosity and efforts of all those involved, although the authors accept responsibility for any inaccuracies that may have occurred. We would like to express sincere gratitude to all contributors. Our friends, John M. and Donna Campbell, introduced us to Peter Schiffer and loaned us numerous photographs from their archives and hours of time in the darkroom. The following collectors, veterans, and historians were major contributors: Dale Edwards, Joe Maguire, Michael J. Perry, Mick Prodger, Burt Sheriff, and Wayne "Donnie" Watts.

Several businesses and other organizations made significant contributions of material: Jeff Spielberg of JS Industries, Inc., P.O. Box 5178, Santa Monica, CA 90409; Mike Conner of The Hobby Shop, Crossroads Mall, Oklahoma City, OK; Manion's International Auction House, Inc., Box 12214, Kansas City, KS 66112; Silent Wings Museum, Terrell, TX; The 45th Infantry Division Museum, Oklahoma City, OK; Edward H. White II Memorial Museum, "Hanger Nine," Brooks Air Force Base, San Antonio, TX: and Mike Davidson, The Hussar, Tulsa, OK. Special thanks to J. Michael Nordin for advice and counseling.

Other contributors and helpers were: Willis Allen, Claude Berry, Bob Biondi, Charles W. Blount, Earnest J. Butler, Robert E. Burnham, Martin Callahan, Tom Carmichael, Tommy Colones, John L. Conway, Fernando Cortez, Gerald Cullumber, Dave Delich, Cyril Dworak, Lonnie J. Eggleston, Jeff Ethell, Dolph Farrand, Richard Fort, Robert Glaeser, Dave Goleman, Larry Gostin, Bob Hammack, Mark Hatchel, Arthur Hayes, P.A. Hermman, "Pete" Houck, Paul Howell, Mark Huntzinger, Jeff Huston, Fred Huston, Ralph Jenks, Jamie Jones, Lt. Col. Rocklin D. "Rocky" Lyons, U.S.M.C., (Ret.), Nesbit L. Martin, Jim McDuff, Robert J. Meer, Randy Mitchell, Eric Mombeek, Pat Moran, William Morris, James W. Mirick, Daryl O'Rear, Gregg Parlin, Richard Peacher, Clay Perkins. Harold Ruth, Jack Rector, Paula Shurtz, Tim Smetana, A.J. Smetana, Troy Shaw, Howard Sossamon, Jim Spaw, Charles Stansel, Dave Summers, Scott Thomas, "Edge" Thomas, Joseph Tucker, Mike Whitson, Ron Willis, Charlie Wilhite, and Gary Valant. Also thanks to Mel Kemph A.V.G.

A very special thanks goes to Rhonda Maguire, who typed my part of the manuscript, and to both of our wives, Rhonda and Martha, who put up with hours of late night phone conversations and time away from our families.

The authors would enjoy hearing from those with flight jackets or other militaria who wish to assist in future projects. Please write to them courtesy of the publisher at the address given below.

On the dust jacket
Lt. C.A. "Dolph" Farrand wearing his A-2 in front of a B-25H of the 490th Bomb Squadron. See page 88 for details. (C.A. Farrand)

Page 1
Lt. R.L. Baggett, 402nd Fighter Squadron, 370th Fighter Group. See page 147.

Front end sheet:
American airmen pose in front of a DH4, circa 1918.

Rear end sheet:
Air Force Captains Richard M. Pascoe (left), and Norman E. Wells from the 555th Tactical Fighter Squadron, add another star to an F-4C Phantom after claiming their second MiG kill on June 5, 1967.

Book Design by Robert Biondi.

Second Edition
Copyright © 2000 by Jon A. Maguire and John P. Conway.
Library of Congress Catalog Number: 93-84498.

All rights reserved. No part of this work may be reproduced or used in any forms or by any means – graphic, electronic or mechanical, including photocopying or information storage and retrieval systems – without written permission from the copyright holder.

"Schiffer," "Schiffer Publishing Ltd. & Design," and the "Design of pen and ink well" are registered trademarks of Schiffer Publishing, Ltd.

Printed in China.
ISBN: 0-7643-1065-8

We are interested in hearing from authors with book ideas on military topics.

Published by Schiffer Publishing Ltd.
4880 Lower Valley Road
Atglen, PA 19310 USA
Phone: (610) 593-1777
FAX: (610) 593-2002
E-mail: Schifferbk@aol.com.
Visit our web site at: www.schifferbooks.com
Please write for a free catalog.
This book may be purchased from the publisher.
Please include $3.95 postage.
Try your bookstore first.

In Europe, Schiffer books are distributed by:
Bushwood Books
6 Marksbury Ave.
Kew Gardens
Surrey TW9 4JF
England
Phone: 44 (0)208 392-8585
FAX: 44 (0)208 392-9876
E-mail: Bushwd@aol.com.
Free postage in the UK. Europe: air mail at cost.
Try your bookstore first.

Contents

Foreword 7
Introduction 8

Chapter I   The American Flight Jacket: History & Development 11

Chapter II   China-Burma-India, & the Pacific Theatre 63

Chapter III Europe, Africa & the Mediterranean Theatre 115

Chapter IV   Unidentified Theatre & Others 197

Chapter V   Navy & Marine Corps Jackets 209

Chapter VI   Korea to Desert Storm 231

Joe D. Maguire, pilot, 321st Air Transport Squadron, 27th Air Transport Group. Photo was taken in Paris, Christmas 1944.

Charles W. "Chuck" Blount, glider pilot, 61st Squadron, 314th Troop Carrier Group. Photo was taken in London, September 1944.

Dedicated to my dad, Joe D. Maguire, and to a good friend, Chuck Blount.

FOREWORD

The adornment of aircraft reached unparalleled levels during the Second World War. The drab repetitive rows of machines that left the aircraft factories in the United States did not remain so after their arrival in the theatres of war. Leggy ladies, home states, animals, caricatures, and every type of lucky symbol all became subjects of the aircraft nose-artist's brush. Couple these artistic representations with the inevitable "name" that airmen gave their aircraft and each airplane became a distinctly personal thing. Once personalized, the aircraft would never be simply a tool to the men who flew it; the aircraft and the fliers would ever after share an identity.

Paralleling aircraft nose-art during the war was its direct by-product, painted flight jackets. For want of a better term, this "jacket-art" became as popular as nose-art. As I knew a bit about nose art but initially very little about jacket-art, I was somewhat at a loss about addressing this unique art form in this foreword. Given the obvious relationship between the two, however, it is apparent that the same desire for individuality that prompted a man to decorate his airplane would encourage him to decorate his jacket. Humans strive for individuality, and this striving for an identity is keenest under circumstances when it is discouraged, as in a war. Every man is uniformly clothed, every man is a part of a team, but every man wants to distinguish himself from his peers in some way. What better way to set oneself apart, yet still be a team member, than to wear some distinctive art on one's jacket? The jackets conferred individuality as well as pride in service to the wearer. Some commanding officers frowned on the practice, but a majority seemed to tolerate it, which brings us to this book. The authors have amassed a spectacular array of jackets within these pages, illustrating what can only be termed a particularly American type of folk art. When possible, the authors have explored the man who wore the jacket as well, probably the most important aspect of collecting wartime memorabilia. The jackets and their original owners have obviously aged since they rode together into battle fifty years ago. Enemies have reconciled, memories have dimmed, and new wars have been fought, but no force in history could ever diminish the deeds of those young men of a half-century ago who answered their country's call. Long after the last flyer has gone west, the jackets will remain as a memorial to those brave men and the world they fought to save.

John Campbell

Aviation author, photographer, and archivist, John M. Campbell, as a U.S. Navy flight officer with VP-26, February, 1970.

INTRODUCTION

From the time I was a small boy I remember looking in the hall closet with awe at my father's World War II uniforms. I thought they were the greatest things I had ever seen, with gold "U.S." and winged propellers on the lapels, winged star insignia on the shoulder, and silver wings on the left chest, which signified my dad was a pilot. There was a box of photographs in the kitchen cabinet taken in far away places – England, France, Belgium, and Germany – of my dad next to his airplane wearing an A-2 jacket and his "50 mission crusher" hat. Dad wore out his A-2 jacket working as a plumber after the war, and his hat was lost long before I was born. However, those uniforms and photographs are proudly displayed in my office today!

This book is written from the perspective of a historian, a collector, and of a little boy grown up who loves and admires his dad and his dad's contemporaries, and the regalia that remains from the war they fought – and won. Only after becoming an adult could I truly understand and appreciate the sacrifice veterans have made and the debt we owe them. All too often, collectors of militaria become engrossed in the "item," forgetting the history behind it, the individual who originally owned it, proudly wore it, and possibly died for it. It is appalling to think of how many historical groupings have been split up and the items sold separately to "maximize the profits" of militaria dealers. The value of a significant number of jackets herein has been enhanced by the historical documentation and additional material accompanying them.

Military collectibles are a tangible link to the past. In the case of American flight jackets, they have been over Germany, Japan, France, England, the Pacific, China, Korea, Vietnam, Iraq, etc. They have been in the cockpits of airplanes that legends are made of – Spad, Flying Fortress, Liberator, Skytrain, Lightning, Thunderbolt, Mustang, Avenger, Dauntless, Hellcat, Corsair, Shooting Star, Phan-

Introduction

tom, and Eagle are but a few. Most importantly, they have been on the backs of American airmen who have fought to keep America and the world free.

Part of the reason for writing this book is to preserve and document the history of as many authentic American flight jackets and the individuals who wore them as possible. Every effort has been made to insure that each jacket in this work is authentic. If any reproductions or fakes have slipped through, we sincerely apologize. We have also made every effort to identify unit insignia, but in some cases, we simply were not able to do so. We welcome help from the reader to identify unidentified pieces.

Unfortunately, in any collectable field, when prices escalate, the unscrupulous start producing counterfeits. We have chosen not to address this issue because when you tell the counterfeiters what they are doing wrong, they correct it and get better. This work will hopefully provide the reader with an excellent representative collection of authentic jackets for comparison. We have also not attempted to provide a price guide. Flight jackets are worth what someone is willing to pay on a given day. From my perspective, each one is a treasure, which brings us to another reason for this work. Since it would be impossible for an individual to own all of these "one of a kind" pieces, we can at least own and enjoy photographs of them.

John Conway and I have been friends for a number of years and have chosen to work with and help each other improve our knowledge and collections, as opposed to competing, as all too often happens in this arena. We hope this book will be a valuable addition to your library and a help to you in expanding your knowledge and/or improving your collections. We welcome any comments and are constantly seek new information and photographs.

Jon Maguire

CHAPTER I

The American Flight Jacket: History & Development

Perhaps no other garment has been so readily associated with a profession since the days of knights in armor, as the flight jacket. The first aviators adopted leather as their material of choice because of its durability, warmth, and, of course, good looks! Although cloth jackets have always been around and used extensively, the leather jacket immediately comes to mind when one hears "flight jacket." Perhaps the pinnacle of leather jackets was reached with the A-2 flight jacket made famous by U.S. Army airmen of World War II. World War II veterans speak of these garments with reverence, and the fact that numerous examples exist after fifty years is a testimony to their durability. These jackets also made a wonderful canvas for self expression. American soldiers have always been individuals – especially airmen! A-2 jackets were painted with everything from pin-up girls to airplanes. They were also decorated with unit patches and nicknames. U.S. Navy and Marine Corp flyers have equal affection for the G-1 jacket. Although seldom encountered with paintings, G-1s are often adorned with numerous patches.

From the earliest days of flight one of the first challenges faced by airmen was that of keeping warm at altitude. For this reason flight jackets have been a major part of aviation from the start. The problem of keeping warm reached its peak in the early days of World War II when large numbers of airmen went aloft in unpressurized, unheated bombers. These missions were flown at altitudes over 20,000 feet where temperatures could reach 60 degrees below zero. This, coupled with open waist windows and drafty nose turrets, made conditions miserable. The first high altitude gear was made of shearling. The classic shearling jacket of that period is the B-3. As the war progressed, electric suits were developed and less expensive, more easily supplied, cloth jackets began to replace the leather gear. The B-10 and B-15 were widely used and well liked by American airmen. These jackets are also found with paintings, although not as commonly as A-2s. In the years following World War II synthetic fabrics were developed. Fire resistant agents became available, and leather faded in utility, but not in popularity. The Navy kept the G-1 in its inventory and, in recent years, the Air Force has reissued the A-2.

Through contemporary and modern photographs, this chapter attempts to show the progression of American flight jackets from the privately purchased coats of World War I to the high-tech garments worn in Desert Storm.

Period photo of Lt. Lewis Rabe wearing a leather flying coat very similar to the example worn by Sydney Noel (top page 12). Rabe was also a member of the 148th Aero Squadron. This American squadron was compiled of American airmen who were trained by the British Royal Flying Corps in Canada and flew British aircraft overseas. Lt. Rabe is resting against the wing of a Sopwith Camel. (Army Signal Corps)

OPPOSITE: Lt. Holten of the 77th Pursuit Squadron wearing an A-1 jacket, circa early 1930s. (Edwards)

WORLD WAR I AND THE "GOLDEN AGE" OF FLIGHT

This full length leather flying coat was used by Sydney Noel of the 148th Aero Squadron. It is actually a British issue specimen and still retains a 1913 dated nomenclature tag inside. The light areas in color are due to uneven lighting at the time the photo was taken. Many examples of flight garments are still in the possession of the veteran or family and must be studied when and where they are found! Note the side entry map pocket on the left chest (detail below). This feature was almost never overlooked by producers of this type of garment.

Right: Captain Jacques A. Swaab poses casually against the wing of a captured German Fokker D VIII in November of 1918. He wears an unusually short leather flying coat that may be of European origin. Captain Swaab was credited with ten aerial victories in the war. (Moses)

The American Flight Jacket: History & Development

A canvas WWI aviator's coat with large fur collar. Note the map pocket on the left chest. A well made garment, it was probably more durable than leather and no doubt much more comfortable. (45th Infantry Division Museum)

Rear view of the WWI canvas aviator's coat. (45th Infantry Division Museum)

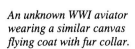

An unknown WWI aviator wearing a similar canvas flying coat with fur collar.

Left: 1st Lt. Harmon Rorisow wearing a canvas aviator's coat with the 1st army "A" sleeve patch sewn to the left shoulder. Lt. Rorisow won the Distinguished Service Cross for scoring 3 aerial victories in 1 sortie over Beaumont, France, November 3, 1918. His Spad aircraft bears the shooting star insignia of the 22nd Aero Squadron. (Brooks via Moses)

Right: Capt. Ray Bridgeman, 22nd Aero Squadron, circa 1918 in a dark canvas flying coat. Bridgeman was credited with 4 aerial victories and served with the Lafayette Escadrille prior to service with the U.S. Army. (Brooks via Moses)

Period photo of a group of WWI army aviation personnel posed outside an aircraft hanger. There are several variations of canvas flying coats visible, in addition to other regulation and non-standard items of clothing.

The American Flight Jacket: History & Development

Above: Insulated canvas flying coat worn by bombing military aviator Lt. Edward Poler. This example has a heavy fleece collar and has been embellished with the single braid rank trefoil of a 2nd Lt. on the cuffs. Below: Front view of Lt. Ed Poler's insulated canvas flying coat. Note the long slash pockets located deep in the sides and the simple button loops that would allow the wearer to adjust it for bulky clothing worn underneath.

The army type A-1 leather flight jacket introduced in November of 1927. This garment was intended for wear as a summer flight jacket for pilots and was the predecessor for the famous A-2 jacket that would follow in just 4 years.

Photo taken April 27, 1930 of army Major Clagette wearing the A-1 leather flying jacket at United airport in Burbank, California.

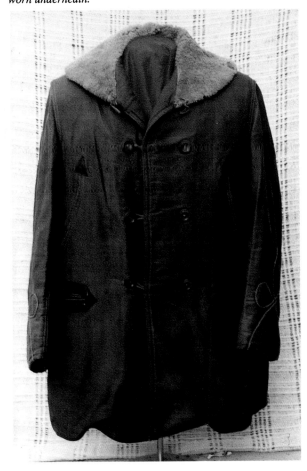

Wool version of the type A-1 flight jacket. Heavy and well made, this example bears a well worn nomenclature label but has held up very well. It is almost identical in cut and design to the leather version.

Period photo of a group of army aviators in the late 1920s. Note the WWI style flight suits. Of special interest is the lone A-1 worn by man 3rd from the left in the front row. Right: Detail of the army aviator wearing the only A-1 jacket present in the above photo. The soft texture of the leather is evident in the left sleeve.

Early photo of a group of Marine Corps aviators in a mixture of cloth and leather jackets similar to the army type A-1.

Close up of three Marine aviators taken from above photo showing clear examples of the cloth and leather variations of their A-1 style jackets.

The American Flight Jacket: History & Development

1920-1930s army aviators posing on the flight line in an interesting variety of garments.

Detail of the three men in the center of the front row of the above photo. Note the A-1 summer flight jackets worn by the men on the right and left, also the Air Corps regulation M-1926 officer's service coat worn by man in center. This coat was also an authorized flight garment. Regulations stated the insignia worn on it would be of an embroidered construction so it would not snag in the parachute harness.

Interesting example of a serge wool flight jacket used by an army air cadet in the 1930s. Although pocketless, it appears to be patterned after the A-2.

THE A-2 JACKET

Left: Pre-World War II army aviator's studio portrait. Note the tattered cloth streamer on his flying helmet. Also evidence of the name tag having been removed from the left chest of his A-2 flight jacket. The jacket may have been a photographer's wardrobe item or it may have been reissued as it was not classified as a garment for permanent issue.

Above: Specifications required that the A-2 jacket be constructed of horsehide and lined in spun silk. The A-2 was service tested in 1930 and adopted as standard issue in 1931. It was limited to issue of existing stocks in April of 1943 and replaced by the AN-J-3. This example is a WWII period production specimen and is lined in cotton. It retains sewn leather lieutenants rank insignia, impressed leather name tag and AAF decal insignia on the shoulder.

Detail of the collar closure hook and press stud fastener for right side of collar. These features are unique to the A-2 flying jacket.

Right: Detail photo of the "Army Air Forces" patch design sometimes applied by manufacturers of the A-2 during WWII. The insignia is actually a water transfer decal that is applied directly to the surface of the leather.

The American Flight Jacket: History & Development

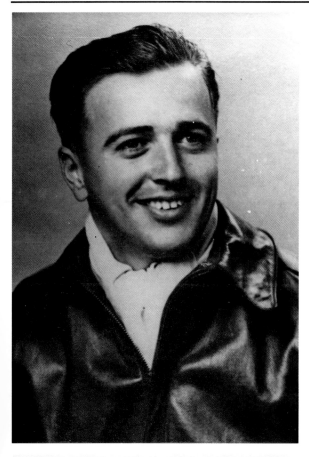

Left: Flight officer "Pete" Houck, fresh out of flight school and posed in his brand new A-2 jacket. Houck was a glider pilot in the 94th Troop Carrier Squadron and was captured by German forces upon landing in Bastogne on December 27th, 1944. He would spend the rest of the war as a guest of the Germans at Stalag Luft 1. (Houck)

Detail photo of a woven A-2 specification label sewn to lining inside the neck. Aero leather was a noted manufacturer of the A-2 during WWII.

Left: An Army Air Force officer in the China-Burma-India theatre in an A-2 flight jacket. Note the two vertical rows of light stitching on each side of zipper. This was a method of sewing cloth or leather "blood chits" inside the jacket. This served a dual purpose: 1. The colorful chits were concealed and, 2. when left unstitched across the top they formed beautiful interior pockets! (John Campbell)

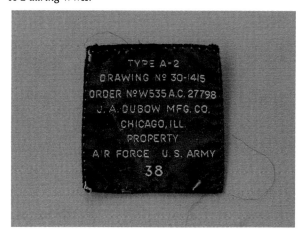

Detail photo of a woven A-2 specification label that has long since parted with its garment. J.A. Dubow was a manufacturer noted for quality.

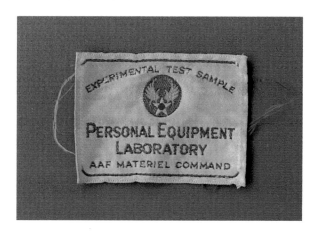

Rarely encountered label for an experimental test garment released through the personal equipment laboratory of the Army Air Forces Material Command.

A beautiful hand embroidered silk "blood chit" of the type often sewn inside the A-2 jacket. The Chinese message promised anyone a reward for assisting the bearer back to allied lines.

Right: A very large example of the A-2 jacket worn by General Berman as he is being welcomed upon arrival via an Air Service Command C-47.

Right: Army Air Force fighter pilot "Dixie" Jackson (right) wearing a simply adorned A-2 jacket. (John Campbell)

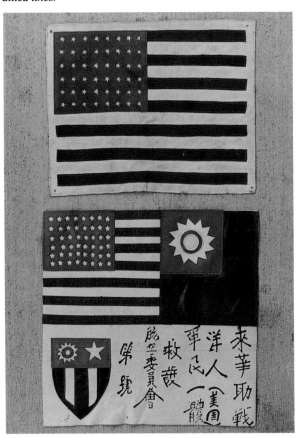
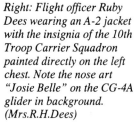

Two theatre produced "blood chits" constructed of multi-piece leather. Chits of this type were often worn on the back of flying jackets but it was thought they would make good targets should the aviator ever be sighted on the ground after a bail out or forced landing.

Right: Flight officer Ruby Dees wearing an A-2 jacket with the insignia of the 10th Troop Carrier Squadron painted directly on the left chest. Note the nose art "Josie Belle" on the CG-4A glider in background. (Mrs.R.H.Dees)

The American Flight Jacket: History & Development

Hand painted leather squadron insignia of the 10th Troop Carrier Squadron – the same design as painted on flight officer Dees' A-2 jacket. (Robin)

The original hand painted leather "Thermal Sniffers" patch that once adorned the jacket of flight officer Browning. (Browning)

Flight officer Louis Browning (2nd from left) with squadron mates pose in front of the first CG-4A glider to be assembled and flown overseas. Note the unofficial "Thermal Sniffers" dumbo insignia painted on the nose. (Browning)

Hand painted leather sleeve patches for the 15th and 8th Air Forces. These were popular items for "dressing up" the A-2 jacket and the production of these types of insignia was wide spread anywhere there was an air base. Local craftsmen in England, Italy, and the China-Burma-India theatres thrived on the income generated from sales of these insignia and would often trade them for American cigarettes or chocolate bars.

Flight officer Louis Browning of the 37th Troop Carrier Squadron standing along side downed plane somewhere in Tunisia. Note the large squadron patch and leather name tag sewn to the left chest of his A-2. Although difficult to see, it is the unofficial "Thermal Sniffers" insignia worn by glider pilots in the unit. (Browning)

Italian made 85th Bomb Squadron patch. The design is tooled on leather and then painted by hand. Insignia of this type was common for units operating in the Mediterranean during WWII and many of them came home sewn on A-2 jackets.

Right: This dozing fly boy sports the ominous looking squadron insignia of the 402nd Fighter Squadron, 370th Fighter Group, 9th Air Force painted directly on the chest of his A-2. Note the "Army Air Forces" decal on the left shoulder. Other articles of interest include his distinctive "crusher" cap, "Crash" (identification) bracelet on his right wrist and type A-11 "Hack" watch on his left wrist. (John Campbell)

This rather nonchalant P-51 pilot wears a likeness of a former aircraft painted on the chest of his A-2. Often personalized insignia depicting aircraft or individual motifs was worn in place of squadron insignia. (John Campbell)

Flight officer Robert J. Meer at Lipa, Philippines, wearing the "Glider Wolf" insignia of the 1st Glider Provisional Group painted directly on his A-2 jacket. (Meer)

Right: Fighter pilot Joe Kuhn with classic "50 Mission Crusher" cap and A-2 jacket sporting an odd size name tag and an applied leather squadron patch on the left chest. (John Campbell)

Far left: Flight officer Jack Bates, somewhere in Sicily, 1944 with hand painted wing and 314th Troop Carrier group insignia on his A-2 jacket. (Hare)

Left: Striking a relaxed stance, this officer has bulging pockets and a large leather squadron patch on his A-2 jacket (John Campbell)

The American Flight Jacket: History & Development

Flight officer Troy Shaw, glider pilot of the 1st Air Commando Group posing for the photographer in Rangoon, Burma. He wears the insignia of the glider section with his name painted above in English. The large, rectangular patch on his right chest has his name painted in native languages. Note also the partial image of a C.B.I. (China-Burma-India) patch on his right shoulder. (Shaw)

The comfort and popularity of the A-2 jacket is readily evident in this photo of two airmen returning from a long range escort mission. (John Campbell)

This Army Air Force tail gunner wears two impressed leather name tags on his A-2. The upper denotes his crew position – "Tail Gunner" and the second his name – "Wagner." This type of personalization was not unusual.

Lieutenant Joe D. Maguire of the 321st Air Transport Squadron at Le Bourget, France, January 1944. Note the tan leather name tag sewn to the left chest of his A-2.

Note the issue pattern tan leather name tag on the left chest of this A-2 jacket. This was a specified pattern that was issued for wear, usually six to a man. (John Campbell)

Interesting photo clearly showing an example of non-metallic officer's rank insignia being worn on the A-2 jacket – lieutenant's bar on the left shoulder. (John Campbell)

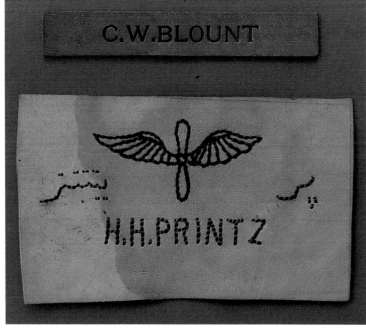

An impressed gold leaf name tag still sewn to an A-2. Lt. Draghi was a group lead navigator in the 15th Air Force.

Issue pattern impressed leather name tag (top) for Charles W. Blount – 61st Troop Carrier Squadron – and a foreign made "Private Purchase" specimen with man's name in Arabic and English and a simple winged propeller motif to denote Army Air Force affiliation.

Right: Note the tan leather, issue pattern name tags and leatherette major's leaf on these A-2 jackets. (John Campbell)

Below: Interesting example of a non-regulation name tag with impressed air gunner wing design and name. Variations in these are endless due to the thriving market at hand for craftsman stateside and overseas.

Officer's of the 23rd Fighter Group pose alongside one of the group's famous "Shark Mouth" P-40s. Officer on the left wears an A-2 jacket with a non-metallic pilot wing, issue pattern name tag and 74th Fighter Squadron patch on the left chest. (John Campbell)

The American Flight Jacket: History & Development

Examples of thin leather officer's rank insignia as issued and worn throughout World War II.

Group of 20th Fighter Group pilots wearing the 77th Fighter Squadron insignia on their A-2 jackets. Man 2nd from the left is wearing the lined cotton B-10 jacket. Some pilots preferred the cloth jackets for comfort and warmth. (John Campbell)

Insignia denoting general affiliation was sometimes worn, even though unofficial. This "PX" patch can be seen on one jacket in the "Holly Terror III" B 17 crew photo.

The entire crew of the B-17 "Holly Terror III" in A-2 leather jackets. Note the limited ornamentation. The man in the lower right wears a "United States Air Corps" patch with aircraft insignia star design on his jacket. (John Campbell)

Crew of the B-24 "Miss N-U" wearing A-2 jackets in a variety of decorated and unadorned configurations. This could be a reflection of individual tastes, but more likely an indication of what was available to who and when. Replacement crewmen or even entire crews would often miss the chance to wear a squadron patch because of enforced regulations or late assignment to their unit.

This 23rd Fighter Group insignia was hand embroidered in china on a 10 x 12 inch section of white silk.

Pilots of the 23rd Fighter Group pose with two of their famous P-40 planes. Several A-2 jackets can be seen and a fairly wide range of colors can be noted in the leather. Three of the men wear the distinctive 23rd Fighter Group insignia. (John Campbell)

This colorful patch is a beautiful rendition of the insignia worn by the 23rd Fighter Group – heirs to Flying Tiger fame. Note that this example appears to be the same type worn in the group photograph.

The American Flight Jacket: History & Development

Captain Sam Trave of the 347th Fighter Group wears a very dark shade A-2 jacket with the little silver "Good Luck" bell attached to the closure hook on the collar. The little bells were a popular souvenir for airmen in the Mediterranean. Examples have been noted with war dates and inscriptions from Rome, the Isle of Capri, San Michele and simple "Good Luck Companion" motifs. (Trave)

Detail of a good luck souvenir bell attached to the collar closure hook on a 12th Air Force attributed A-2 jacket. The bell bears the Italian inscription: "La Campania de la Fortuna – Roma."

A variety of good luck souvenir bells in different sizes, metals and bearing different inscriptions.

John T. Godfrey – 336th Fighter Squadron, 4th Fighter Group alongside his P-47D "Reggies Reply." Note the police style survival kit whistle attached to the collar closure hook of his A-2 jacket. The whistle came as an element of various air crew survival kits and was intended to provide a means of making noise if forced down at sea. They became a popular ornament or zipper pull on flight jackets during the war and seemed to have been most popular in the E.T.O. (John Campbell)

Detail of the air crew survival kit whistle still attached to the collar closure hook of Fred Christensen's A-2 jacket. A fighter "Ace," Christensen scored 21 1/2 kills over German aircraft including a six in one day total. There is a good photo of Christensen wearing this jacket, with whistle attached, in the WWII 56th Fighter Group unit history.

Close up of an A.A.F. aerial bomb safety pin tag used as a zipper pull on an A-2 jacket. Although a bit chic at the time, one can well imagine how handy it could be for crewman wearing gloves or mittens in flight.

Close up of a British R.A.F. issue air crew whistle with issue markings impressed on the side. Many of the British issue examples were packed in U.S. kits and it is not unusual to find them on American issue jackets.

The AN-J-3 jacket was basically the Navy G-1 jacket manufactured without a shearling collar. It was introduced in April of 1943 and intended to replace the A-2. It was apparently procured in limited quantities and is very seldom encountered. One example encountered did have a shearling collar and was issued through USN supply channels.

Rear view of the AN-J-3 leather jacket showing details of the pleated bi-swing back and the sewn-in belt across the back. These details had previously been confined to the USN G-1 jacket.

Right: With restrictions of issue, shortages, delays and final discontinuation of A-2 jackets during the war it became necessary for some airmen to procure them through other channels. This example was tailor made in Sydney, Australia and is a very accurate and detailed rendition of the issue item. Some examples of foreign made, issue pattern A-2 jackets have also been noted, apparently manufactured in compliance with lend-lease agreements.

Below: Detail of the commercial Australian maker's label in a privately procured A-2 jacket.

WORLD WAR II SHEARLING JACKETS: THE D-1 AND B-3

Sergeant Robert D. McCall, glider mechanic, on the flight line with his trusty pipe and D-1 jacket. (McCall)

A D-1 jacket worn by an Army aviation cadet.

The shearling D-1 jacket was intended for use by ground crew in cold climates, but many of them found their way into the air.

A noted B-26 with two of her ground crewmen. Note the D-1 jacket worn by man on the left.

An unusual variant of the D-1 jacket – this example has no exterior pockets. This soldier displays an unusual uniform of the day!

Distinguished pilot, Jack Ilfrey, beside his P-51 "Happy Jack's Go Buggy." His ground crew accompany him and three D-1 jackets can be seen in the photo. (John Campbell)

Another pocketless variation. The stocking cap worn by this smiling crewman was also an Army Air Force issue item for ground crewmen.

D-1 jackets worn by ground crewmen accompanying the pilot of an early P-51. (John Campbell)

The B-3 jacket was adopted as the cold weather flight jacket by the Army Air Corps in 1934. It was noted for its warmth and durability.

The American Flight Jacket: History & Development

The multi-hide construction of the B-3 jacket provided outstanding quality due to careful matching of the best material at hand.

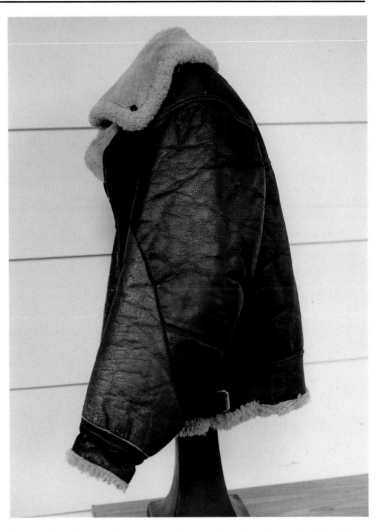

A smooth leather overlay on wear points, such as the sleeves, is evident in the photo. Note also the small tab which was sewn into the shoulder seam on each side. This was apparently for rank insignia.

Sgt. Herman Hetzel of the 458th Bomb Group, 753rd Squadron standing in front of one of the units impressive lion head B-24s. Note the single, angular patch pocket on the lower right front of his B-3 jacket. The single pocket was a distinct feature of the B-3 jacket. (John Campbell)

An entire B-24 crew of the 458th Bomb Group wear B-3 jackets. (John Campbell)

Nice period photo of army airmen in B-3 jackets and the A-3 trousers designed to complete the cold weather flying outfit.

This air crewman wears an Army Air Force A-2 jacket under his B-3. The "layer" principle for warmth in clothing was essential to comfort at high altitudes.

This waist gunner wears the B-3 jacket and the rest of the accompanying shearling items to complete a high altitude gunner's wardrobe. (John Campbell)

A pilot with the 367th Fighter Group wears the B-3 jacket while resting against an impressive war souvenir! (John Campbell)

The American Flight Jacket: History & Development

An Air Cadet, stateside, wearing the B-6 flying jacket and A-5 trousers designed for intermediate temperatures. These were probably very welcome when flying open cockpit trainers.

The AN-J-4 jacket was introduced in May of 1943. It was an improved model designed to replace the B-3. Short in cut-it was thought to be more practical for closed cabin flying than the B-3. The "AN" prefix indicated the item could be carried in both army and navy inventory.

Woven specification label sewn inside the left front interior of the AN-J-4 jacket.

Right: The C-3 shearling vest was introduced in 1936 and was thought by many to have been adopted for use by air gunners due to its non-restrictive effect on the arms and shoulders. It was actually an optional garment designed as an under layer for other jackets.

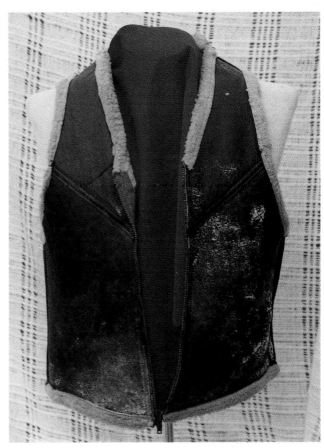

WORLD WAR II CLOTH JACKETS: THE B-10 & B-15

Detail showing the very dark OD cotton material and stitched on epaulet of the type A-9 jacket.

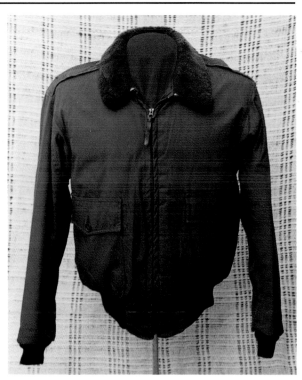

The AAF issue type A-9 jacket was apparently a predecessor of the B-10 jacket but little is known about it.

Glider pilots of the 61st Troop Carrier Squadron a few days before the "Varsity" mission across the Rhine in March of 1945. Left to right they are: Lt. Gene Canova (in B-10 jacket), Flight Officer Charles W. Blount (wearing B-9 jacket) and Lt. Frederick D. Manget (in armor crewman's field jacket). Flight Officer Blount was the only one of the three to survive the mission. (C.W. Blount)

The American Flight Jacket: History & Development

The type B-10 AAF flying jacket was classified as an intermediate weight winter garment. It could be worn as a middle layer garment under heavy jackets and over shirts. Constructed of light O.D. shade cotton, it was alpaca lined. Below: Detail of the small, ink stamped AAF property logo stamped on the left shoulder of a B-10 jacket.

B-10 flying jacket shipped home with the effects of Captain John B. Eaves of the 56th Fighter Group. Captain Eaves was killed in a flying accident in England. He was credited with five kills at the time of his death.

Flight Officer Robert J. Meer wears his B-10 jacket with a large, embroidered felt 67th Troop Carrier patch sewn to left chest. Meer stands near a memorial somewhere in occupied Japan. Meer was a participant on Operation Gypsy- the last airborne operation of WWII into Appari, Luzon. (Meer)

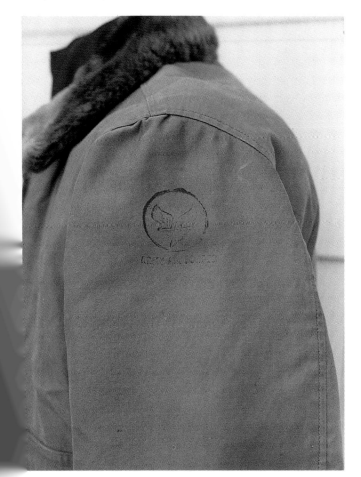

Glider Pilot Flight Officer Russell D. Parks wearing the type B-10 flying jacket.

Australian made 67th Troop Carrier patch worn by Flight Officer Robert J. Meer. This patch is the same one worn on Meer's B-10 jacket in the period photo shown in this section.

P-40 pilot with leather name tag sewn to the left chest of his B-10 flying jacket.

Lieutenants Lee Powers and Joe D. Maguire of the 321st Squadron, 27th Air Transport Group, wearing B-10 jackets at Grove, England –1944.

Unnamed Top Turret Gunner astride a Martin B-26 "Marauder" with barrels of the turrets' twin .50s in hand. There is a faint drawn design on the left chest of his B-10 jacket.

The American Flight Jacket: History & Development

This flyer eagerly awaits hot coffee (note canteen cup in his right hip pocket). The pencil pocket in leading edge of left pocket is evident on this B-10 jacket.

Right: Pilots of the 20th Fighter Group inspecting a captured German aircraft. Note all but one of the flyers wears the B-10 jacket. The kneeling man in the center wears the army issue M-1941 field jacket. (John Campbell)

Right: Three army airmen apparently somewhere in the CBI theatre. The two men on the left wear shearling B-3 jackets while the man on the right wears a B-10 with the China-Burma-India theatre patch on his left shoulder. (John Campbell)

P-38 pilot with the 367th Fighter Squadron. Note survival kit "Police" style whistle hooked to zipper pull. (Campbell)

Right: Unusual example of a flying jacket designed specifically for female pilots, the B-16. Obviously based on the B-15, this jacket is not known to exist in any modified form. Right below: The printed specification label inside the WASP issue B-16 flight jacket. Note AAF inspector's counter stamp in upper left.

Left: The B-15 jacket was considered an improved version over the B-10 and became standard issue in April of 1944. This is the first style B-15 with no modifications. The B-15 jacket went through many variations of manufacture and is the basis of today's MA-1 flight jacket. Below left: Detail of the white ink printed AAF logo on the left shoulder of this early B-15 jacket. Note also that there is not a pocket on the sleeve below the insignia, simply provision for storage of four writing instruments. Many of these early B-15s did not even have a specification label inside!

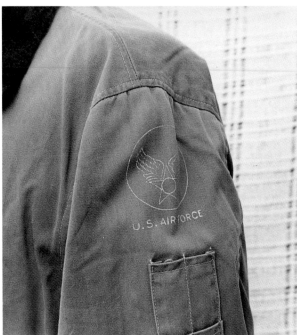

Right: Late-war version of the B-15A with leather tabs added to chest for oxygen mask clip and small button loops near armpits for wire looms. An interesting aspect of the later versions construction is the zipper is offset to the wearer's right about two inches.

Left: Flight Officer Richard C. Cochran, P-51 pilot with the 384th Fighter Squadron wearing the B-15 jacket with invasion flag patch on the sleeve. The American flag was a universal form of identification and could save an airman's life when forced down in combat. Note the 94th Bomb Group B-17 in the background.

NON-STANDARD & VARIANT CLOTH JACKETS

The army issue armor crewman's winter field jacket was popular in virtually every capacity of cold weather use. It was the first expedient "winter" flying jacket used by the Eighth Air Force and was immensely popular with fighter pilots due to limited cockpit space. Constructed of light olive drab cotton, it was blanket wool lined.

Lieutenant Sam Fine wearing the army armor crewman's field jacket in the cockpit of a CG-4A combat glider. (Fine)

Flight Officer Doug "Bear" Baly wearing the armor crewman's jacket with invasion flag armband, over a pair of herring bone twill coveralls. This photo was taken shortly before Baly's departure for southern France as the pilot of a CG-4A glider. (Sutton)

The M-1942 airborne troops parachute jump jacket. Constructed of light olive green shade cotton twill, this jacket was unlined but featured four billow type pockets. With the matching belt, it gave the appearance of a bush type jacket and was very popular with aircrew who were fortunate enough to obtain them.

A group of 1st Air Commando pilots gathered in front of their thatch hut "Bashas" in Burma. Note man on the right in back row wearing the parachute jump jacket. The Air Commandos involved a rather elite group of airmen in specialized operations in Burma. Several of the pilots were issued the parachute jump jacket and trousers while undergoing tactical training in North Carolina –just before departure for Burma. They may have been the only group of airmen to obtain these garments through authorized supply channels. Note film star Jackie Coogan, in the lower left. (Scott)

Custom made summer weight flight jacket tailored somewhere in China for a fourteenth Air Force Fighter Pilot. This is a non-standard pattern based on a civilian design. Constructed of cotton, it has slash pockets and a zipper front. The owner's name, "Nielson," is hand embroidered in Chinese characters on the left chest.

Dark olive drab shade gabardine jacket designed for wear over the electrically heated F-3 jacket. This material denotes officer's usage.

Officer's gabardine over-jacket for the F-3 heated jacket. Note its resemblance to the Eisenhower jacket and the lack of the fur collar.

Over-jacket for the F-3 heated suit constructed of coarse olive drab serge wool for use by enlisted ranks.

The American Flight Jacket: History & Development

The type K-1 summer flight jacket issued to Army Flight Nurses. Adopted in January of 1945, this jacket was constructed of unlined, tan cotton.

The type L-1 flight jacket was designed for use by flight nurses and was adopted by the AAF in January of 1945. Constructed of dark olive drab gabardine, this jacket was rayon lined and not unlike the nurse's Eisenhower service jacket.

The British army "battle-dress" jacket was first issued to U.S. aircrews while training in North Africa for the famous Ploesti mission of August, 1943. This came about as a remedy to the short supply of uniforms for U.S. troops in the area. This was just the beginning of the popular "Ike" jacket issued to U.S. troops from WWII to 1957.

Detail of the "United States" nationality title and various insignia found on this British battle dress jacket.

Lieutenant Robert Price, glider pilot with the 96th Troop Carrier Squadron wearing a British battle dress jacket adorned with insignia. (Price)

Two Army Air Force glider pilots somewhere in North Africa in early 1943. Note man on left wearing a British issue battle dress jacket with his wing badge on the left chest. The man on the right wears an early M-1938 field jacket. (Robert Wilson)

Army Air Force officer's B-13 flight jacket. An issue item, this jacket was carefully researched and designed to be the ultimate combination dress/flight jacket. Wearing metallic insignia on it was not encouraged due to the fact that it might tangle in parachute harnesses, etc.

An unusual combination of service jacket and flight jacket, this British contract ETO jacket is constructed of dark OD shade Barathea wool.

The American Flight Jacket: History & Development

The enlisted man's version of the British "lend-lease" contract ETO jacket. Originally intended as an outer garment for ground personnel, many of these were worn as a "layering" garment of outer garment in flight. These exist in at least three distinct varieties.

The Army Air Force issue B-14 flight jacket for enlisted men. Basically the same as the officer's B-13, the B-14 is different only in the shade and weave of the wool of which it was constructed. Right: Detail of the B-14 flight jacket showing slash pockets and optional insignia.

Group of 422nd Night Fighter Squadron pilots at Chateaudun, France in August of 1944. Note man third from left wearing a B-13 officer's flight jacket. (John Campbell)

Three P-61 ground crewmen wearing British contract ETO "field" jackets. Note patch and slash pocket varieties on left and center. (John Campbell)

WORLD WAR II PARKAS: THE B-9 & B-11

The type B-11 "jacket" was a parka type garment, fabricated of cotton with an alpaca lining. Standardized in 1943, it was one of the earliest cold weather garments to see service in WWII. Below: The B-9 was essentially the same as the B-11 but featured a quilted lining rather than the alpaca lining.

Captain C.E. Weaver with his P-51 in the background. The puffy sleeves of his B-9 jacket indicate the down/feather fill was adequate for a cold weather, high altitude garment. (John Campbell)

1st Lt. Irvin C. Kinney wearing the B-9 jacket in the glider pilot's "Tent City" in Saltby, England. Lt. Kinney was one of nineteen American glider pilots to survive the ill-fated "Ladbroke" mission into Sicily, July 9, 1943.

C-47 pilots Flight Officer Robert Bailey and Lt. Lee Powers of the 321st Air Transport Squadron, 27th Air Transport Group at Grove, England, 1944. Both wear the AAF type B-9 jacket but note the light and dark shades of lambswool collars.

This photo shows detail of the quilted rayon lining used to add cleanliness and ease of donning to the B-9 jacket. (John Campbell)

The American Flight Jacket: History & Development

ARMY FIELD JACKETS USED BY AIRMEN

A group of U.S.A.A.F. officers including General Claire L. Chennault (center) standing in front of a B-25. The first three men from the left wear M-41 field jackets, while the remaining four men wear A-2 leather flying jackets.

The Army issue M-41 field jacket was an extremely popular garment during the war and seemed to have been available to airmen, judging by the many examples encountered and visible in photos. It was constructed of light weight olive green cotton and lined in a shirt weight OD serge wool. Noted for comfort, it was not a durable garment and specimens in good condition are quite unusual today.

Army quartermaster specification label found inside the pocket of an M-41 field jacket.

Detail of a leather First Lieutenant's bar sewn to the shoulder of an M-1943 field jacket.

Army issue M-1943 field jacket constructed of dark shade, olive drab cotton. Designed to be worn as an outer garment or as a shell for other layers of clothing. These were quite popular with airmen throughout the war but they were actually a mainstay with glider pilots due to their ability to function as ground combatants once they landed.

Flight Officer Richard K. Fort, a glider pilot with the 91st Troop Carrier Squadron, wearing the M-41 field jacket. Fort wears complete foot soldier's combat rig and is awaiting departure on Operation Dragoon – the Invasion of southern France. (Fort)

Lieutenant James W. Campbell of the 79th Troop Carrier Squadron. Jim wears the M-1943 field jacket and trousers. Jim is awaiting departure on the "Varsity" mission to Wessell, Germany in March, 1945. (Campbell)

Glider pilots Jim Kennedy and Bennett Y. Allen preparing for the "Varsity" mission in March of 1945. Allen was a major and led an element on this operation. Both men wear the M-1943 field jacket. (Allen)

An embroidered felt 158th Liason Squadron patch.

Right: A jovial member of the 158th Liason Squadron wearing an unusual M-1943 jacket. This jacket has been cut down to waist length in "Ike" jacket fashion. Insignia of the squadron has been sewn to the chest pocket. A 9th Air Force patch can be seen on the left shoulder.

The American Flight Jacket: History & Development

U.S. NAVY & MARINE CORPS JACKETS

The USN G-1 leather flying jacket is still a regulation naval item. This is a nice example of a WWII contract G-1. The G-1 was originally designated the M-442. It has remained basically unchanged since its inception in the 1920s.

Right: Detail of the "Bi-swing" pleat in the shoulders of the USN G-1 jacket. This, along with the button flap pockets, fur collar, and belted back, differentiated the G-1 from the A-2 in cut and design. The original contract specified that the G-1 would be constructed of goatskin, whereas the A-2 was of horsehide.

Left: This WWII Navy pilot exudes the spirit of the naval aviator and his fabled flying jacket. The stance is very indicative of the comfort the G-1 could provide.

Right: Interesting array of uniforms with two G-1 jackets worn by Marine aircrewmen of a JM-2.

These naval airmen wear G-1 leather jackets, one adorned with over-sized torpedo squadron patch (left). The man in the center wears the light weight cotton flying jacket designated type M-421A under his life vest.

An unusual haircut worn by a Navy aircrewman wearing a G-1 jacket. A PB4Y-2 (Naval version of the B-24) can be seen in the background.

Two marine aircrewmen in G-1 jackets with a JM-2 aircraft painted in yellow for tow target duty. The JM-2 was the naval version of the B-26.

The entire crew of the USN PB4Y-2 "Little Joe." Note traces of insignia on the left chests of several of their G-1 jackets.

A.V.G. "Flying Tiger" crew chief Walter Dolan in China circa 1941. Note Dolan's unadorned G-1 jacket. Walter Dolan passed away in January of 1983. (Mel Kemph)

The American Flight Jacket: History & Development

A less noted WWII vintage naval flying jacket was the fleece lined M-445A. Like the army counterpart, type B-3, this jacket was issued with accompanying shearling trousers, type M-446A.

Right: WWII vintage USN "G-1" jacket with the M-422A designation on specification label.

This olive green cotton flying coat was alpaca lined and designated as the upper half of a winter flying suit. Some examples have been noted with documented coast guard usage. (Phillips)

A variety of leather name tags ranging from WWII to the 1970s.

WWII vintage USN "G-1" jacket with scarce AN-J-3A designation on specification label. The "AN" prefix suggests that this version was available in both army and navy inventories. Period photos document the use of what appears to be G-1 style jackets by army aircrew but little is known about this phase of possible issue.

Vietnam era USN "G-1" jacket with the last and most universally used designation for this jacket. G-1 on the specification label.

U.S. AIR FORCE FLIGHT JACKETS: 1947 TO DESERT STORM

Liason pilots of the 7th division aviation section pose for the camera on a snowy runway somewhere in Korea. Note all but one of the pilots wear a WWII vintage B-15 flight jacket. Lieutenant Claude A. Berry (left) is a three war veteran of army aviation, having flown gliders in WWII, liason planes in Korea and helicopters in Vietnam. (Berry)

Lieutenant Norman "Duffy" Dufresne simulates cockpit activities for the photographer at Randolph Field in February 1950. Dufresne wears an early P-1 pattern jet pilot's helmet and a WWII vintage A-2 jacket. Dufresne was with the first group of fighter pilots to be called up for Korea. A graduate of flight school in 1949, Dufresne managed to acquire an A-2 through non-issue channels in hopes of looking like an old timer on base. The A-2 enjoyed fairly widespread usage well after the government stocks had been exhausted. (Dufresne)

Left: The B-15B was the first post-WWII production of the B-15 pattern jacket and was the basis for the modern MA-1. The B-15B differed from the WWII B-15A only in that it was constructed of nylon rather than cotton. (JS Industries)

Lieutenant "Duffy" Dufresne flew 125 missions in P-80s over Korea and won the DFC and air medal. Dufresne was photographed in Korea wearing an Army Air Force WWII issue B-15 jacket. WWII vintage flight clothing carried with the laurels of a seasoned aviator and was considered quite prestigious among jet pilots. (Dufresne)

The American Flight Jacket: History & Development

The B-15C was the first of the innovative new blue USAF flight clothing. Other than color, it was identical to the B-15B.

The B-15D was again much like its WWII counterpart but introduced the long lasting "Sage" green nylon to the U.S. Air Force.

This group of Korean war flyers wear an interesting mixture of the old and new in flight gear. From left to right: A WWII vintage B-15 jacket with dark blue L-2 flight suit; a B-15C jacket (dark blue nylon) with a WWII vintage L-2 flight suit in olive drab; an early USAF N-2 parka in OD nylon with a pair of WWII vintage A-11 insulated cotton trousers; a WWII vintage B-15 jacket with unidentified, insulated flight suit. Mixed periods and colored in flight clothing was very typical of the Korean war.

Left: Interesting shot of Korean war B-26 crewman wearing the B-15C jacket. The high gloss of the nylon gives the jacket a leather-like appearance.

Above: The B-15D-modified was the transition between the B-15 and the MA-1 style jackets. The B-15D was modified by removing the fur collar and replacing it with a short knit one. Below: Detail of some basic construction details of the B-15D-modified that reveal the unchanged basic design from the WWII B-15A.

Detail photo showing the "modified" label sewn right over the original B-15D label in the neck. It is not known if the modifications were executed through Air Force channels or if the work was contracted, but the quality and positioning of the label might suggest the work of private contract labor.

The first pattern L-2 flight jacket was constructed of olive green nylon and differed from the B-15B in degree of insulation, style of collar and addition of epaulets. (JS Industries) Below: Woven specification label sewn in the neck of the 1st pattern USAF L-2 jacket. Note the faint USAF winged star ink stamp directly below. (JS Industries)

The American Flight Jacket: History & Development

The L-2A jacket in dark blue nylon.

1950s USAF pilot wearing the L-2A flight jacket and holding a P-4 style flight helmet. Note placement of the unit insignia and plastic covered name tag on the chest. (John Campbell)

Lt. Colonel Ralph Durnbaugh of the 465th Tactical Fighter Squadron wearing an L-2A flight jacket as he inspects his aircraft prior to take off. (John Campbell)

Detail photo of metallic press stud fasteners installed on the left chest of the L-2A jacket. This type of work was usually executed on a local command basis. This was done to make the owner's name tag detachable.

An example of the "Light Zone" label used in the L-2A jacket.

Another variant of the "Light Zone" label for the L-2A jacket. (JS Industries)

The L-2B jacket in sage green nylon. Note the green satin lining.

Detail photo of woven specification label sewn in the neck of an L-2B. Note detailed analysis of material used in construction along lower edge.

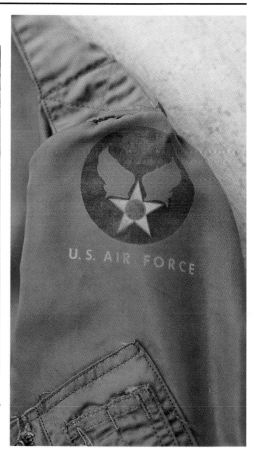

Color printed "U.S. Air Force" logo on the left shoulder of an L-2B jacket.

Below: U.S.A.F. pilots wearing flight clothing during a Squadron Level Award ceremony. Man on the left is wearing an MA-1 insulated jacket. The men behind and to the left wear L-2B jackets. All appear to be sage green. (John Campbell)

The American Flight Jacket: History & Development

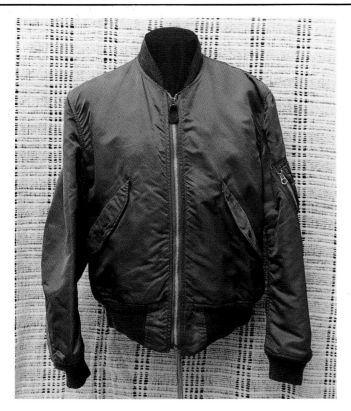

A late version of the L-2B jacket in sage green nylon but made without the press stud fastener flap on waist band. This version had a printed label inside the pocket and was lined in bright "indigo" orange.

Late version of the L-2B reversed to show the indigo orange lining which was intended to aid visibility on the ground in the event of bail-out or forced landing.

A direct descendant of the WWII B-15 jacket, the MA-1 was well insulated and comfortable. Like the L-2B, the MA-1 eventually was lined in indigo orange as a survival aid. Note the nylon web oxygen mask clip strap on the chest and the printed U.S.A.F. logo on the left shoulder.

Printed specification label found in the left pocket of late pattern MA-1 jackets. Note traces of the orange lining in background.

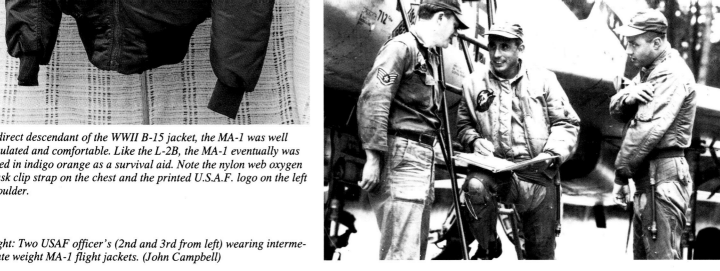

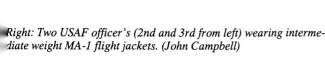

Right: Two USAF officer's (2nd and 3rd from left) wearing intermediate weight MA-1 flight jackets. (John Campbell)

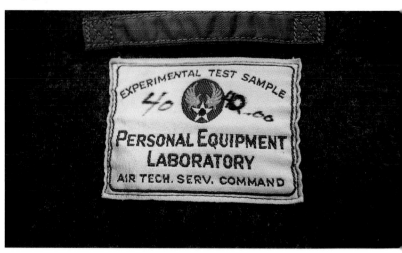

Earliest version of the type A-1 wool flying shirt. Constructed of olive drab wool, this jacket style shirt was introduced during WWII but this specimen bears a USAF "Experimental Test Sample" label. Below: Detail of the test sample label sewn in the A-1 flight shirt.

Late version of the olive drab wool A-1 shirt with early USAF issue label in neck.

The A-1A was the second pattern in the series. Introduced in the early 1950s, this garment worked efficiently in the layering method for warmth.

The American Flight Jacket: History & Development

The last pattern of this style garment was the A-1B constructed of sage green wool. All patterns of the A-1 shirt were accompanied in issue by matching E-1 trousers.

Above: The N-2B flying parka in sage green nylon. (JS Industries)

Below: Detail photo of the woven specification label inside an N-2B parka. (JS Industries)

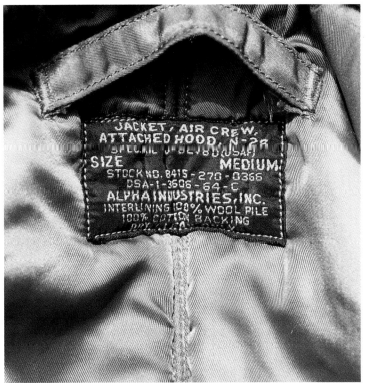

The N-2A flying parka was well insulated and fitted with a fixed, alpaca lined hood trimmed in wolf fur. The N-2 version that preceded it was basically the same, but constructed of olive green nylon.

The N-3 flying parka was longer in cut than the N-2 and was designed for Arctic flying conditions. The N-3 was constructed of olive green nylon and featured a shearling lined hood trimmed in wolf fur. (JS Industries)

Late pattern N-3B parka in sage green nylon. (JS Industries)

Detail photo of the woven specification label inside an N-3 parka. (JS Industries)

Detail photo of the woven specification label inside the neck of the N-3B flying parka. (JS Industries)

The American Flight Jacket: History & Development

The N-3A flying parka in dark blue nylon.

Detail photo of the color printed USAF logo on the left shoulder of the N-3A parka. (JS Industries)

Right: Detail photo of the woven specification label inside the neck of an N-3A aircrew parka. (JS Industries)

Assortment of impressed leather and sealed plastic officer's rank insignia as worn on USAF flying garments.

CHAPTER II

China-Burma-India & the Pacific Theatre

On December 7, 1941, the harsh reality of war was brought home to the American people with the Japanese victory at Pearl Harbor. The early days of the Pacific war progressed badly for the United States with a string of losses and withdrawals extending from the Philippines, Java ıd New Guinea. The Japanese advance was finally stopped with their tempted invasion of Port Moresby and the Battle of the Coral Sea.

U.S. Naval and Marine forces played a major role in the U.S. victory the Pacific. Names like Midway, Guadalcanal, and Tarawa bring to ind legendary conflicts which resulted in hard fought American ctories against seemingly impossible odds.

Aerial victory in the Pacific was much more of a team effort between e Army Air Force and Naval and Marine forces than in other theatres operation. The Army Air Forces involved in the Pacific war were the h, 7th, 11th, 13th, and 20th.

The 5th Air Force was active in both offensive and defensive actions the Southwest Pacific from September of 1942 under Lt. General eorge C. Kenney. The 5th was engaged until January 1943 using all of cir capabilities to defeat the Japanese from Papuan and New Guinea. ıis included shipping strikes, defense of allied installations, troop and ıpply transportation and air/ground cooperation. In March of 1943 the h destroyed a Japanese convoy in the Bismarck Sea and cooperated in nding on New Guinea and New Britain. During the same period the h developed bombing techniques which proved very successful in tacks on Wewak and Rabaul.

On June 15, 1944, the 5th was consolidated with the 13th Air Force as e Far East Air Forces moved to coordinate the air drive toward Tokyo. ıe 5th Air Force produced America's all time "Ace of Aces," Richard Ira ong, with 40 kills.

The 13th Air Force was active in the South Pacific with its major task ing to halt Japanese advances in the Solomons. In the Army Air Forces ntribution to the Guadalcanal campaign P-39s were the work horses ground support and the P-38s gave the bombers long-range escort in riking bases in the Solomons.

After the capture of Guadalcanal in February of 1943, the Allied rces pressed northwest to New Georgia, Bougainville, the Admiral- !s, Rabaul and Kavieng. The Carolinas also were within range of the th's B-24s. After Truk and Wolea were in allied hands, the 13th moved est from the admiralties to assist the 5th Air Force in hitting Noem-Foor

Cloth American flag and leather blood chit on the back of Captain Forrest J. Bell's A-2 (see pages 70-71).

OPPOSITE: 490th Bomb Group B-25H (see page 88). (Dolph Farrand)

Detail of Eugene Hammack's A-2 (see page 73). (Hammack)

Detail of H.F. Schaub's A-2 (see page 74). (JS Industries)

Island in preparation for MacArthur's assault. The 13th also was involved in air actions which contributed to the fall of Saipan and the invasion of the Philippines. As part of the Far East Air Forces, the 13th and 5th Air Forces were focused on coordinating the air offensive to Tokyo.

The 7th Air Force operated in the Central Pacific and devoted early efforts to defending the Hawaiian Islands. B-17s and B-26s of the 7th saw action at Midway and from July to February 1942 the 11th Bomb Group was engaged in the aerial offense of the Solomons. Lack of supplies plagued the 7th until mid-1943 when B-24s of the VII Bomber Command were prepared for the offensive in the Central Pacific.

The 7th was primarily engaged in hitting enemy bases as a prelude to amphibious assaults. Strikes were flown against Tarawa, Makin, Nauru, Wotje, Jalvit Maloelap, and Kwajalein. The fall of Tarawa allowed for establishment of air fields used for operations against the Marshalls.

As the campaign moved west, targets included the Carolines, Truk, Ponape, and the June assault on Saipan. In August of 1944 the 7th was reorganized as a Mobil Tactical Air Force in preparation for the drive to Tokyo.

The North Pacific and defense of Alaska were the responsibility of the 11th Air Force. The Aleutian Islands were not only a vulnerable target, but they also provided a base of operations for offensive attacks on the enemy. The 11th played a key role in stopping the Japanese assault on Dutch Harbor, Attu and Kiska.

These actions allowed the 11th to hit targets close to the Japanese homeland, including the Kurile Island. In 1944 and 1945, B-24s and B-25s hit targets on Shimushu and Paramushira which held the strongest concentration of Japanese air and ground forces on the islands. All actions in the North Pacific were carried out in the face of continually bad weather. The activities of the 11th forced the Japanese to deploy valuable air power to the North.

The 20th Air Force took the war to the Japanese homeland with the very-long-range, very heavy B-29 Superfortress. The first raid flown by the 20th was on Bangkok, June 5, 1944. Ten days later, about fifty B-29s hit the steel center of Yawata on Kyushu for the first real blow to the Japanese homeland from bases on Japan. The B-29s began a concerted assault on Japan's aircraft industry. By the end of this assault about 6 Japanese cities were virtually destroyed and over 175 square miles of Japan was in ruins. The 20th struck the final blows of World War II with the dropping of atomic bombs on Hiroshima and Nagasaki. To end the war, the Japanese surrendered unconditionally on August 14, 1945.

After Pearl Harbor the first significant Naval Air battle occurred May 6th through May 8th, in what became known as the Battle of the Coral Sea. This was the first Naval battle in which the participants never saw their enemy. Aircraft carriers launched attacks at each other over the horizon. At the Coral Sea the Japanese suffered their first set back. There can be little doubt that with the loss of the aircraft carrier Lexington, destroyer Simms and the Neosho, the United States suffered a tactical loss. In the exchange the Japanese lost one light carrier, the Shoho, and one destroyer. The strategic effect of the battle however was to stop the Japanese advance southeastward and their planned landing at Port Moresby in New Guinea.

The battle of Midway occurred between June 4 and June 6, 1942 and has been called by many the most decisive Naval battle of all time. Midway changed the course of the Pacific war when in a matter of hours the Japanese lost four aircraft carriers, the Kaga, Soryu, Akagi and

Horyu. In exchange the United States lost the aircraft carrier Yorktown and a destroyer. The aircraft losses in this battle were 147 U.S. aircraft to 332 Japanese. Perhaps the most important outcome of this battle was the loss of many of the first line fighter pilots in the Japanese Navy, a loss which Japan was never able to make up.

On August 7, 1942 U.S. forces in the Pacific took the offensive in the invasion of the Solomon Islands. The First Marine Division reinforced by elements of the Second, landed on Guadalcanal, Talagi, Gavutu and Tananboga. Between August, 1942 and February, 1943 the Solomon campaign raged. Fighter units of the Army Air Corps supplemented and assisted Marine and Navy units throughout this campaign. The combined air forces operating off of Guadalcanal acquired the name "Cactus Air Force."

With the completion of the Solomon campaigns the major Naval activities moved to the central Pacific with the battles to capture or neutralize the Gilbert and Marshal Islands. These campaigns saw extensive use of Naval and Marine carrier units in the close air support function. Marines pioneered the use of aircraft for close air support during the occupation of Haiti during the 1920s.

During late February, 1944 the United States Task Force 58, under the command of Vice Admiral Raymond Spruance, commenced operations against the Mariannas Islands. While preparing to land troops on Saipan, Tinian, Rota, and Guam, Naval and air units engaged in hundreds of strikes against the Islands as well as participating in what has become known as "The Mariannas Turkey Shoot," perhaps the most one sided aerial battle in history. Until the end of Japanese resistance on Guam on August 10, 1944, Naval and Marine air units continued to provide close air support and fighter cover throughout the Mariannas.

The most significant result of the successful Mariannas campaign was the capture of Tinian and Saipan which could be used for bases for the B-29 superfortress for strategic raids on the home islands of Japan.

On September 15, 1944 the Marines landed on Pelelieu and for a month, while the First Marine Division fought one of the bitterest land battles of the Pacific war, air support and air cover was provided by the carriers.

Throughout October, 1944 task force 38 under command of Admiral William "Bull" Halsey attacked Rabaul, Formosa and the Philippines. In one week of October, 1944, Naval forces destroyed 73 Japanese ships and 670 Japanese aircraft.

During the week of October 23 through 26, 1944 the largest Naval battle in American history took place the region of the Philippine Islands. The battle of Leyte Gulf resulted in the loss of 34 Japanese Naval ships including 3 battleships, 4 aircraft carriers, 10 cruisers, 13 destroyers and 5 submarines. Many of these ships were lost to Naval and Marine aircraft units working in conjunction with U.S. submarines.

Iwo Jima was attacked on February 19, 1944 by United States Marines. Until the invasion, Japanese fighters had been using the island as a base to intercept U.S. bombers attacking the home islands. While taking the Island of Iwo Jima was costly in terms of human life, once captured the island provided a necessary emergency landing field for B-29 crews returning damaged from missions from Japan's home islands. Between February and August of 1945 several thousand airmen owed their lives to the availability of Iwo Jima as an emergency landing field.

In the last and the bloodiest amphibious operation of WWII Naval aviation met one of its severest challenges. Japanese kamikaze attacks which had first been seen during the battle of the Philippines, became a daily occurrence. Throughout the campaign for Okinawa, Naval and

Detail of an unnamed 1st Air Commando A-2 (see page 79). (Michael J. Perry)

Detail of Tex Hill's G-1 jacket (see page 82).

Detail of E.F. Pearsall's A-2 (see page 83). (Richard Peacher)

Marine air units continued to act in the close air support role as well as providing fighter cover for the largest fleet ever assembled. On April 6, 1945 the first concentrated Kamikaze attack of over 350 suicide planes struck. U.S. fighters and Naval gun fire succeeded in downing all but 24 of the suicide planes which went on to hit their targets resulting in great destruction including damage to the battleship North Carolina. The following day Naval air units of task force 58 located and sank the super battleship Yamato and one cruiser, 4 Japanese destroyers and 54 Japanese aircraft at the cost of 10 U.S. aircraft downed. Between April, 1945 and the end of the war on August 15, Naval aviation continued to pound Japanese targets in the home islands as well as what remained of Japan's empire.

The combined efforts of the Army air force and Naval and Marine air units resulted in an unbroken string of victories from June of 1942 until the end of the war. The joint effort of the various branches of the service can perhaps best be exemplified by the mission of the Enola Gay on August 6, 1945. While the world remembers that Lt. Colonel Paul Tibitts was the aircraft commander for that mission, most have forgotten that the bombardier and mission commander was Navy Captain William "Deak" Parsons. The victory in the Pacific was truly a joint effort of land, sea, and air forces which could not have been accomplished without the contributions of each of the services.

THE CHINA-BURMA-INDIA THEATRE OF OPERATIONS

The China-Burma-India Theatre of Operations was characterized by isolation, both geographically and through enemy actions. The three countries of China, Burma and India were grouped together by U.S. planners simply for strategic convenience, since most of Burma was lost to the Japanese fairly early in World War II and China was essentially surrounded throughout the war. China was undoubtedly the most impoverished theatre of operations in which the United States fought, with supplies, fuel, rations and spares scarce throughout the conflict. The CBI area of operations is unique in that it can be viewed as essentially an aerial theatre of war. In the China-Burma-India Theatre, much more than anywhere else in the war, aviation actions far outweighed ground troop actions for most of World War II; besides the U.S. Army Air Forces, British and Empire troops contributed mightily to the ultimate Allied success there. Two widely separated U.S. Air Forces operated in the CBI: the 10th Air Force, with headquarters in Delhi, India, and the 14th Air Force, headquartered in China.

The 10th Air Force was positioned in India to prevent a link-up between Axis forces driving east through North Africa and west through Burma. They undertook bombing and escort missions into Burma, China, and Thailand, protected the Hump, and coordinated with the Air Transport Command in supplying the isolated 14th Air Force in China.

The 14th Air Force was the descendant of two prior entities. The first was the American Volunteer Group (A.V.G.), popularly known as the "Flying Tigers," the second was the China Air Task Force (CATF). The A.V.G. was composed of former U.S. Army, Navy and Marine flyers and ground crewmen who were recruited as "civilian" advisors and technicians with Chiang Kai-Shek's Chinese Air Force. Once the U.S. formally entered the war, the A.V.G. actively opposed the Japanese. The A.V.G. was dissolved in July 1942 after seven months of combat. Succeeding the A.V.G. was the CATF which was a stopgap group of military pilots that took up the slack after the departure of the Tigers and before the creation

Leather American/Chinese Blood Chit on the back of W.G. Mumford's A-2 jacket (see page 72).

China-Burma-India & the Pacific Theatre

of the 14th Air Force. The 14th Air Force, despite constant problems with spare parts and supplies, undertook an unceasing defense of the Hump and an aggressive campaign of ground and shipping attacks, ranging all over China.

The Flying Tiger's legacy lived on, after a fashion, in the 14th Air Force's winged tiger insignia, although it was totally different from the Disney Studios' designed tiger insignia of the AVG. The choice of insignia was understandable since the 14th inherited the Flying Tigers' commanding officer General Claire L. Chennault, as well as certain key personnel of the AVG. The China-Burma-India Theatre of Operations' insignia was a federal-type shield with red and white stripes, a blue field containing a single star and a sun, respectively representing the U.S. and China.

A WORD ABOUT BLOOD CHITS

The term "Blood Chit" will be used throughout the CBI section of this book, so a bit of explanation is necessary. A detailed examination of Blood Chits would be beyond the scope of this work, but a brief word about them may provide a keener appreciation for the photographs in this section.

Blood Chits or "Escape Flags" as they were sometimes called, were made of a number of materials including leather, cotton, silk and rayon. They were designed to provide instant identification to their bearer and serve as a safe-conduct pass with any local people that a downed airman might encounter. Most designs had a flag of the U.S., China, Burma, or the U.K. and a message in several of the region's languages describing the bearer's plight, his desire to be returned to his unit, and the reward to be gained if assistance were rendered. The translation of one American-issue chit reads:

"Dear Friend, I am an Allied fighter, I did not come here to do any harm to you who are my friends. I only want to do harm to the Japanese and chase them away from this country as quickly as possible. If you will assist me, my government will sufficiently reward you when the Japanese are driven away."

This particular chit carried the message in seventeen languages, including Burmese, Chinese, Tamil, Thai, Jawi, Malay, French, and English. Issue-type Blood Chits were serial numbered, accountable items, but privately-made chits fabricated by native craftsmen (and women) could be purchased all over the CBI theatre.

Initially, many airmen wore the Blood Chits on the backs of their jackets, a practice that can be traced to the first U.S. air fighters in the Orient, the American Volunteer Group. As the war ground on, however, Blood Chits became more commonly worn inside the jacket, or sometimes simply carried in a pocket. Several factors contributed to this, such as the discovery by Allied flyers that a prominently displayed Blood Chit made a good target, or that a Nationalist Chinese flag might not have the desired effect on would-be rescuers if the airman was downed in a part of China controlled by the Communists. It is common to see chits moved from the back of an A-2 to the inside, where they did double duty as map pockets.

Early example of a Blood Chit. This one, No. 0042 was issued to A.V.G. flight leader C.J. Rosbert. (JS Industries)

Detail of cloth Blood Chit on Robert Kirk's A-2 jacket (see page 69)

"THE HUMP PILOTS"

One of the greatest coordinated air achievements of World War II was the supply of U.S. and Allied forces over the Himalayas. The Himalayas were collectively called "The Hump" and the men who flew "Over the Hump" were "Hump Pilots." The flight jackets worn by these individuals, operating under stressful and difficult conditions, tend to show a lot of character. Often, painted camels were used as mission indicators on jackets and "blood chits" were almost standard equipment. This effort fell under the Air Transport Command and variations of the ATC insignia, or one of its units, are frequently seen on Hump jackets also.

The first jacket in this section is unnamed, but in the pocket was a wonderful account of a Hump pilot's World War II service. As this record is a fine representative example of a Hump pilot's service, it is reproduced here in full: "If this Jacket had a log book it would say that it had been issued by the U.S. Government to a civilian ferry pilot with the Air Transport Command, Sixth Ferry Group, Long Beach, California, in mid 1943. The Jacket was involved with the ferrying of various aircraft around the country as co-pilot and pilot: BT-13, B-17, C-47, UC-78, B-24, PT-19, AT-6, BC-1, PS-1 B, C & D, P-47, P-38, A-20, L-5.

In early 1944 the Jacket changed from civilian status to pilot officer status and, shortly thereafter, delivered a C 47 to New Dehli, India, and was stationed at the 1326th AAF Base Unit in Lalmanirhat, Bengal, India.

The Jacket then spent seven months instrument flying across India. From Lalmanirhat, the Jacket hauled cargo and passengers to Calcutta and stations in the Assam Valley, from which cargo, passengers, and fuel were hauled into the interior to China.

In November of 1944, the Jacket was transferred to Jorhat, India, and spent a number of months flying the "hump" into the interior of China aboard Consolidated C-87 and C-109 aircraft (converted B-24 bombers). The Jacket was seen in places like Luliang, Chenkung, Kunming, Chengtu, Kwanghan, and Pengshan, China, plus various bases in Burma. The Jacket then went to Gaya, India, where it instructed other jackets in flying four engine tank C-109 type aircraft.

The Air Force and CBI (China, Burma, India) Theatre emblems on the Jacket are government issue. The Chinese flag on the back was standard practice among pilots and was instru-

Above: Unnamed ATC "Hump pilot's" jacket, the pocket of which contained the account at left. Visible, painted on the pocket flap, is the serial number C87-41-24141 and name "Doity Goity" of an airplane on which this individual survived the crash. Note leather ATC patch. Unfortunately, the leather name tag is worn off completely. Also note right front panel with painted camel mission indicators, the backwards camel representing a "turnaround" due to engine trouble. Below: Back of unnamed Hump jacket with leather Chinese flag blood chit. (both - Willis R. Allen)

China-Burma-India & the Pacific Theatre

The three primary aircraft used in supply operations over The Hump were the C-47 Skytrain, the C-46 Commando, and the C-87 "Liberator Express," which was the cargo version of the B-24 Liberator.

Above: Douglas C-47 "Skytrain" in flight. Below: Consolidated C-87 "Liberator Express." (Campbell)

Left: Curtiss C-46 "Commando."

mental in identifying and saving many who had to bail out over the hump. The American flag placed on the inside served the same purpose. Pilots and crews were not assigned to definite aircraft over the hump. However, to a certain extent they were at instructional bases such as Gaya. The specific aircraft identified on the pocket of the Jacket was one from which it was fortunate enough to escape after a landing gear collapse and fire on landing.

The camels represent trips over the hump as first pilot. The one facing the wrong direction represents a "turnaround" on which the Jacket could not complete the flight due to engine problems and had to return to home base.

In September of 1945, the Jacket was discharged from military service and has since been waiting patiently to be of use again.

ROBERT KIRK

Robert Kirk was an ATC pilot. His jacket is a privately purchased A-2 style made by Air Associates, Inc. He purchased it at Love Field in Dallas, Texas. The jacket has a cloth blood chit sewn on the back and a standard issue ATC patch and name strip on the left breast.

A-2 jacket worn by Robert Kirk. (Michael J. Perry)

Back of Kirk's A-2 with cloth blood chit in place. (Michael J. Perry)

Robert Kirk as a civilian instructor. (Kirk via Perry)

W.F. McINTIRE

Lt. Col. W.F. McIntire flew with the India-China Wing of ATC. His A-2 jacket is plain on the outside, except for a leather strip with his name and rank in Chinese and English. The lining of McIntire's jacket has three different blood chits sewn in.

Detail of leather name strip in English and Chinese on McIntire's A-2. (45th Infantry Division Museum)

Below: Interior of Lt. Col. McIntire's jacket with Burmese, U.S. flag, and Chinese chits in place. (45th Infantry Division Museum)

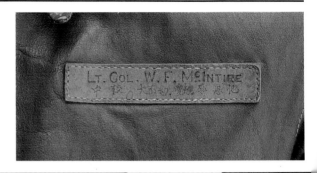

A-2 worn by Lt. Col. W.F. McIntire, India-China Wing ATC. (45th Infantry Division Museum)

FORREST J. BELL

Captain Bell also served with the India-China Wing of ATC. His jacket is truly spectacular with the large variety of insignia it carries.

Right: Front and rear shots of the A-2 jacket named to Captain Forrest J. Bell, India-China Wing, Air Transport Command. Name in English and Chinese on leather, under sterling silver U.S. pilot's wing on left front of Bell's A-2.

China-Burma-India & the Pacific Theatre

Aviation Cadet Club I.D. card with photograph of Forrest J. Bell.

Burmese blood chit/map pocket sewn inside the left front of Bell's A-2.

Multi-piece leather A.A.F. insignia on right shoulder. Also visible is a sterling silver Captain's rank pinned to the epaulet.

Multi-piece leather C.B.I. patch on left shoulder.

Multi-piece leather insignia of India-China Wing, Air Transport Command (ICWATC) on right front of Bell's jacket.

Hand embroidered Chinese blood chit/map pocket inside right front of Bell's jacket.

W.G. MUMFORD

Lt. Mumford's A-2 is another fantastic example of a Hump pilot's jacket. There is a wide variety of insignia, both "official" and novelty. His mission record indicates an amazing 75 trips over the Hump.

Detail of unidentified unit insignia on right front of Mumford's A-2.

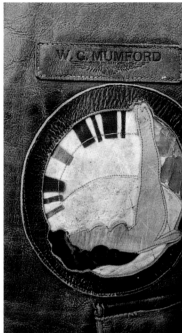

Leather name and wing over novelty version of the ATC patch (which features a reclining nude with her legs raised at a ninety degree angle) on left breast of Mumford's jacket.

A-2 jacket of Lt. W.G. Mumford.

Back of Mumford's jacket. Note camel mission scoreboard representing 75 trips over the Hump, painted on the back of Mumford's A-2. Also, leather American/Chinese blood chit.

The left shoulder of Mumford's jacket has this leather C.B.I. patch sewn in place.

Right shoulder of Mumford's jacket with leather A.A.F. patch.

China-Burma-India & the Pacific Theatre

A.L. TAYLOR

"Rocky" Taylor's jacket is very simple. It has only the novelty version of the ATC patch (as seen on Mumford's jacket), a leather name strip, and the back has "Rocky" in crude letters. The ATC patch is unique in that it is made of aircraft fabric and leather.

A-2 jacket worn by A.L. "Rocky" Taylor, ATC.

Detail of novelty ATC patch constructed of aircraft fabric and leather on Taylor's A-2.

EUGENE E. HAMMACK

Gene Hammack made an incredible 80 Hump trips, as is recorded on the right front panel of his A-2. Also, the name Frances and a painting of the Himalayas is present. Frances is Mrs. Hammack. The inside has a nice leather American/Chinese blood chit/map pocket sewn in. The name tag has his name in three different languages. The back of the jacket is plain.

A-2 worn by Eugene E. Hammack. (Hammack)

Beautifully made leather American/Chinese blood chit used as a map pocket inside Hammack's A-2. (Hammack)

Leather C.B.I. patch on left shoulder of Gene Hammack's A-2. (Hammack)

Leather A.A.F. patch on right shoulder. (Hammack)

H.F. SCHAUB

The heavily adorned A-2 of H.F. Schaub is yet another stunning example of a Hump pilot's jacket. The front left chest has a name strip over a standard issue embroidered ATC patch. The back has the unusual placement of a leather Chinese blood chit over a cloth American flag. Most often they are reversed with the U.S. flag on top. The inside has a Burmese chit.

Back of Schaub's A-2. It is more common to see the U.S. flag over the Chinese blood chit instead of below, as seen on this A-2. (JS Industries)

Detail of leather Chinese blood chit and printed cloth U.S. flag on the back of Schaub's jacket. (JS Industries)

A-2 jacket worn by H.F. Schaub, ATC. (JS Industries)

Detail of Burmese blood chit. (JS Industries)

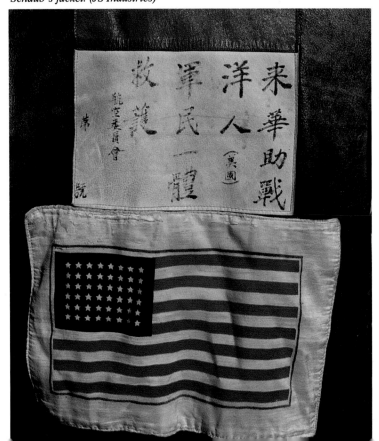

China-Burma-India & the Pacific Theatre

J.A. VITKOVITS

Captain Vitkovits racked up an impressive number of Hump trips, as the camels on the right front panel of his A-2 commemorate. We cannot see under the collar in the photographs, but it appears he made 66 missions. At one time his jacket had a different patch where the ATC patch is now sewn, so he could very well have served in another capacity before becoming a Hump pilot. His jacket also had blood chits on the back which have been removed.

Above: Close up of insignia on Captain Vitkovits' jacket. The ATC patch is a decal on leather, "official" version. The right shoulder has a leather, theatre made, A.A.F. patch and the painted camel, Hump mission scoreboard is clearly visible. The epaulets are complete with leather Captain's bars. (Larry Gostin)

Right: A-2 jacket worn by Hump pilot, J.A. Vitkovits, ATC. (Larry Gostin)

ATC A-2 JACKETS

These two jackets are very simple. The only distinctions are two different official variations of the ATC patch (leather and cloth).

Below: A simple unnamed ATC jacket with fully embroidered ATC insignia. The jacket also has an A.A.F. decal on the left shoulder.

A-2 jacket worn by an ATC airman with leather name tag and decal on leather ATC patch. (JS Industries)

ROBERT E. BURNHAM

Robert E. Burnham retired as a Lieutenant-Colonel from the United States Air Force in 1969. A graduate of Oklahoma A & M, Burnham had a degree in Mechanical Engineering when the war came. Though a freshly minted ROTC second lieutenant of the Army Engineers, he opted for the Air Corps. After training, Burnham's engineering degree prompted his assignment to the Air Materiel Command at Kelly Field in San Antonio. While with the Air Materiel Command he attended an Ohio test pilot school, checking out on a variety of aircraft, including the P-51, P-47, B-17 and B-24.

Burnham was eventually transferred to the Air Transport Command ferrying aircraft. In May, 1945 he reported to Reno, Nevada and drew a C-46 airplane and a new B-15 jacket. He was to ferry the C-46 to a transport unit in Karachi, India, where he expected to be assigned with the aircraft to a Hump unit. After a review of his records showed four-engine qualification, he was assigned to a C-54 unit operating out of Tezgaon in eastern India. Never having flown a C-54 made no difference, so after three or four stints as co-pilot, Burnham became a C-54 pilot and ended the war with 450 hours over the Hump. He was issued the A-2 jacket pictured when he joined the C-54 unit; it had originally belonged to another pilot whose fate is unknown. Blood chits were not worn on this jacket but were carried in the pockets on each Hump flight. Locally-made baseball-type caps were worn by all the airmen in the unit, and can just be seen in the photo below right.

From left, Lieutenants Stealey, Burnham, and Arango at Test Pilot School, Vandalia, Ohio. By-the-book Kelly men Stealey and Burnham still have their cap springs in place while Arango, who came from "hot" Shaw Field, South Carolina, has a 50-mission look to his cap. Burnham's A-2 above was returned to the Air Materiel Command when he transferred to the ATC.

A-2 Jacket of Robert E. Burnham (Dale Edwards)

Multi-piece leather construction CBI sleeve patch

Multi-piece leather construction USAAF sleeve patch

Bob Burnham and his C-54, summer, 1945.

China-Burma-India & the Pacific Theatre

Close-up of the multi-piece leather "Blood Chit" sewn to back of the M-1941 field jacket. Note the addition of the C.B.I. command patch to the design.

A "Hump" Pilot's M-1941 field jacket

The Army quartermaster issue M-1941 field jacket was designed for wear in ground combat in moderate climates. It became one of the most popular G.I. garments of the war and eventually found its way into the air. The China-Burma-India theatre was renowned for rough terrain and a wide variety of atmospheric conditions. This jacket is somewhat restrained in its adornment but conveys the flavor of the theatre of operations and must have been a great deal more comfortable than its leather counterparts in many instances.

Above: Front view of the Army issue M-1941 field jacket worn by a China-Burma-India airman.

Close up of the locally made, printed cotton CBI command patch and stitched leather 1st Lt's bar in the left shoulder area.

Right: Rear view of the CBI M-1941 field jacket. Note the wide waist band, bi-swing shoulders and button adjustment tabs in lower waist area. All these elements plus the comfort of its cotton construction served to make this one of the most popular garments of the war.

C.B.I. Sergeant's M-1941 Field Jacket

This Army issue M-1941 field jacket is carefully adorned with an interesting variety of theatre and stateside produced insignia. The unofficial variety air ferry command patch would seem to imply that governing powers were either reasonably lax, or that this man rarely came in contact with them!

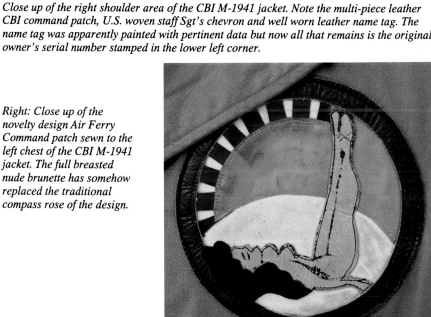

Above: Front view of the AAF CBI Sergeant's M-1941 field jacket. Note the long cuff and buttons for adjustment tabs on the cuffs.

Close up of the right shoulder area of the CBI M-1941 jacket. Note the multi-piece leather CBI command patch, U.S. woven staff Sgt's chevron and well worn leather name tag. The name tag was apparently painted with pertinent data but now all that remains is the original owner's serial number stamped in the lower left corner.

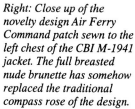

Right: Close up of the novelty design Air Ferry Command patch sewn to the left chest of the CBI M-1941 jacket. The full breasted nude brunette has somehow replaced the traditional compass rose of the design.

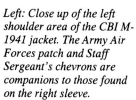

Left: Close up of the left shoulder area of the CBI M-1941 jacket. The Army Air Forces patch and Staff Sergeant's chevrons are companions to those found on the right sleeve.

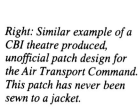

Right: Similar example of a CBI theatre produced, unofficial patch design for the Air Transport Command. This patch has never been sewn to a jacket.

China-Burma-India & the Pacific Theatre

1st AIR COMMANDOS GROUP

ERNIE BUTLER

Flight Officer Ernie Butler served as a glider pilot of the 319th Troop Carrier Squadron of the 1st Air Commando Group in the China, Burma-India Theatre in WWII. The Air Commandos enjoyed a fairly illustrious reputation during the war due to their daring and dangerous exploits in support of Wingate's "Chindits" during the recapture of Burma from the Japanese. Unique also was the fact that their operations were indirectly spotlighted in Milt Caniff's famous comic strip "Terry and the Pirates." The glider pilots of the 1st Air Commando Group were a key element of the operation and were among the first Americans to set foot in the Japanese infested jungles of Burma under the code name of "Operation Thursday." Worn with pride throughout his flying days, Ernie Butler's A-2 jacket was passed down to an admiring sister and two nephews after the war. It still is in excellent condition!

Front view of Ernie Butler's A-2 jacket. Note the absence of large leather name tag from left chest. The tag was removed for post-war use of the jacket by admiring relatives!

Rear view of Ernie Butler's A-2 jacket. Note hand-made, multi-piece leather American flag sewn to the back. The American flag, and similar insignia, were thought to be an essential form of recognition should airmen be forced down in native territory.

UNNAMED 1st AIR COMMANDO GROUP

The second jacket from this unit, although unnamed, is a spectacular piece. It carries a nice variety of different theatre made insignia and a wonderful example of the 319th Troop Carrier Squadron, Glider Section, patch. This insignia is truly one of the most appealing of World War II. It features a smiling mule with wings for ears, superimposed on a one for First Air Commandos, all surrounded by the letter "G" for glider. The mule carries a ghurka knife (kukri) to signify association with the Chindits and the top of the numeral one, on close examination, is a woman's breast.

Left: A-2 jacket worn by a member of the 319th Troop Carrier Squadron, Glider Section, 1st Air Commandos. Note the 319th Troop Carrier Squadron, Glider Section, insignia with the kukri in the mules mouth and the breast at the top of the numeral one. Right: Back of 1st Air Commando A-2 with beautifully made, multi-piece leather American flag. (both - Michael J. Perry)

Leather C.B.I. patch on left shoulder of 1st Air Commando A-2. (Michael J. Perry)

A.A.F. patch, hand embroidered on velvet, on right shoulder. (Michael J. Perry)

Sam Altman, Frank Randall, and Troy Shaw, glider pilots of the 1st Air Command Group, clowning for a photographer in India, 1944.

1st Air Commando glider pilot Nesbit L. Martin shows off the blood chits inside his A-2 jacket.

Jackie Coogan's 1st Air Commandos A-2 on display at the U.S. Air Force Museum.

Actor Jackie Coogan while serving with the 1st Air Commandos, wearing his A-2 jacket.

UNNAMED 13TH BOMB SQUADRON "GRIM REAPERS"

This A-2 jacket is unusual in serveral respects. It has all of the features and characteristics of a standard issue A-2, but the label affixed is civilian. It is difficult to tell if this is what remains of the original label, or if someone through the years has added this where the government label was once attached. The leather name strip appears to have never been embossed with a name. The A-26B Invader "Gunship" painting would indicate late war use, as the 13th flew A-20s during most of the war and did not re-equip with Invaders until 1945 (although they did test the Invader in July 1944). It is also possible the jacket was worn in Korea where the Invader was widely used. The 13th Bomb Squadron was attached to the 93rd Bomb Group, 5th Air Force, and operated from bases in the Pacific including Australia, Mindanao, New Gujinea, Leyte, Mindoro, Okinawa, and Japan. During the Korean conflict the 13th operated from Japan and Kunsan, Korea.

Left: A-2 jacket worn by an airman with the 13th Bomb Squadron "Grim Reapers", 3rd Bomb Group, 5th Air Force. (JS Industries)

everse of "Grim Reapers" A-2 showing A-26B vader painting. (JS Industries)

Detail of blank name strip and 13th Bomb Squadron "Grim Reaper" insignia painted on left chest. (JS Industries)

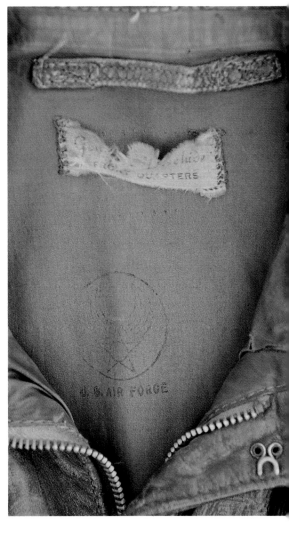

Unusual civilian label sewn on the collar of 13th Bomb Squadron A-2. This label was possibly added later where the government label was once sewn. (JS Industries)

THE 23RD FIGHTER GROUP

The 23rd Fighter Group was formed with experienced pilots and personnel from Chennault's American Volunteer Group and carried on the name of "Flying Tigers." During World War II the 23rd flew from numerous air fields in China using P-40s and P-51s. The World War II squadrons of the 23rd Fighter Group were the 16th, 449th, 74th, 75th, and 76th. The 118th Tactical Recon Squadron was also assigned to the 23rd. The 23rd Fighter Group was assigned to the 14th Air Force.

COLONEL DAVID LEE "TEX" HILL

"Tex" Hill is one of the more prominent figures in U.S. military aviation history. Prior to World War II, he served on the Saratoga in Torpedo Three and aboard the Ranger in Scouting Forty-one. In 1941 he joined the American Volunteer Group ("Flying Tigers") where he served as flight leader and squadron leader of the 2nd Squadron. When the Tigers were disbanded, Tex volunteered to stay in China and took a commission in the U.S. Army Air Corp where he activated the 75th Fighter Squadron, 23rd Fighter Group. He returned to the U.S. to command the Proving Ground Group at Eglin Field in December of 1942 and went back to China in 1943 as commanding officer of the 23rd Fighter Group. Tex again returned to the states in 1944 to command our first jet group, the 412th. Following the war, Hill served in the Air Guard and Reserve program until retiring in 1968. Tex Hill is credited with 18 1/4 victories (12-1/4 with the A.V.G. and 6 with the 14th Air Force). Tex Hill's G-1 jacket is truly a spectacular piece of U.S. aviation history. It has a combination of leather and painted insignia.

G-1 jacket worn by Tex Hill with leather name tag in English and Chinese, and multi-piece leather 23rd Fighter Group patch on left chest. (Hill via Cortez)

Back of Tex Hill's G-1 with painting of a Flying Tiger leaping through the Chinese National insignia, wearing Uncle Sam's hat, and tearing a Japanese flag. Truly a spectacular piece of history! (Hill via Cortez)

David Lee "Tex" Hill, U.S.A.A.F., wearing his G-1 jacket. His flight helmet is a British "C" type and the goggles are U.S. AN-6530s. (Hill via Cortez)

China-Burma-India & the Pacific Theatre

E.F. PEARSALL
Everson Pearsall flew P-51 Mustangs with the 118th Tactical Recon Squadron, 23rd Fighter Group, and was credited with 2 victories. The A-2 jacket worn by Pearsall has all leather insignia.

Left: A-2 jacket, with leather name tag in three languages and multi piece leather 23rd Fighter Group insignia, worn by E.F. Pearsall, 118th Tactical Recon Squadron, 23rd Fighter Group, 14th Air Force. Below: Right shoulder of Pearsall's jacket with leather 14th Air Force patch sewn in place. (both - Richard Peacher)

UNNAMED 23rd FIGHTER GROUP
Regretfully, the name tag is missing from this A-2, but it was worn by an American fighter pilot of the 23rd Fighter Group, 14th Air Force. The leather insignia is very similar to that seen on Hill's and Pearsall's jackets.

Right: Unnamed A-2 jacket, 23rd Fighter Group, 14th Air Force. Above: detail of insignia on unnamed 23rd Fighter Group A-2. (both - JS Industries)

UNNAMED 75th FIGHTER SQUADRON, 23rd FIGHTER GROUP

The name on this A-2 jacket is worn completely off, but the patches show the pilot who wore it served in the 75th Fighter Squadron, 23rd Fighter Group, 14th Air Force. The insignia is theatre-made leather, with the exception of the C.B.I. patch and tattered blood chit, which are cloth.

Unnamed 75th Fighter Squadron, 23rd Fighter Group A-2. (JS Industries)

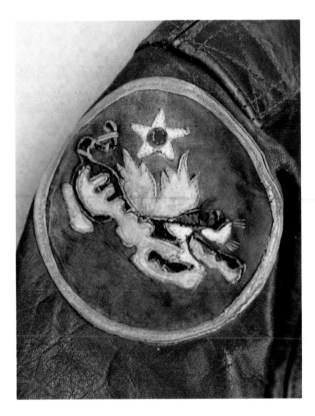

Multi-piece leather 14th Air Force patch on right shoulder. (JS Industries)

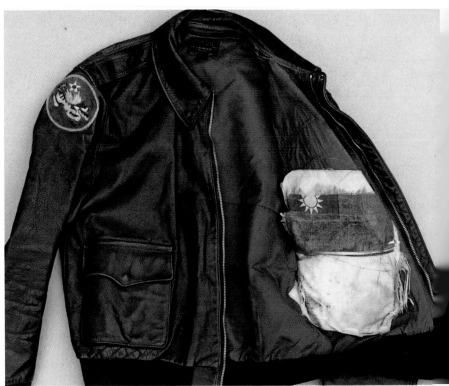

Tattered blood chit inside unnamed 75th Fighter Squadron A-2. (JS Industries)

Right: Multi-piece leather 75th Fighter Squadron patch on left chest. (JS Industries)

China-Burma-India & the Pacific Theatre

822ND BOMB SQUADRON, 38TH BOMB GROUP

The 822nd Bomb Squadron flew heavily armed B-25 Gunships with the black panther insignia covering the nose out of Australia, New Guinea, Biak, Morotai, Luzon, and Okinawa. They were attached to the 38th Bomb Group, 5th Air Force. This A-2 was worn by a pilot with the 822nd Bomb Squadron.

Left: A-2 jacket worn by a B-25 Gunship pilot of the 822nd Bomb Squadron, 38th Bomb Group, 5th Air Force. The black panther patch on the right front is the 822nd. The left front has a pilot's wing on leather, a name strip, and a 5th Air Force patch. (JS Industries)

"BURMA BANSHEES"

The "Burma Banshees" A-2 flight jacket is from the 89th Fighter Squadron, 80th Fighter Group. Its original owner is now unknown. The 80th served with the Tenth Air Force in India and Burma. The paintings were likely done by the "Squadron Artist" and the shoulder insignia locally made in India.

Below left: Front view of A-2 jacket from the 89th Fighter Squadron, 80th Fighter Group, Tenth Air Force. The group insignia is painted on the right breast and heavily worn squadron insignia on the left breast. (Manion's)

Back of 89th Fighter Squadron A-2. (Manion's)

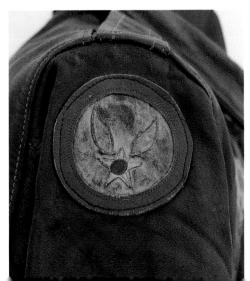

89th Fighter Squadron patch. This is the insignia which was painted on the left front of the "Burma Banshees" A-2.

Far left: Multi-piece construction, leather C.B.I. patch on left shoulder of "Burma Banshees" A-2. Left: Multi-piece construction, leather A.A.F. patch on right shoulder of "Burma Banshees." (both - Manion's)

"THE GENERAL BILLY MITCHELL GROUP"

No information on this A-2 jacket was available. We know from the shoulder insignia that its owner was assigned to the C.B.I. theatre of operations.

Far left: Front/left shot of "The General Billy Mitchell Group" A-2 jacket from the C.B.I. Left: Front/right shot of "General Billy Mitchell Group" A-2. Below: Detail of leather "General Billy Mitchell Group" patch.

EDWARD L. COOK

Ed Cook entered glider training in May of 1942 and graduated as a Sgt. Pilot in August of that same year. After a stint as an instructor state side, Ed was assigned to the 27th Troop Carrier Squadron in the CBI and flew 230 combat missions in C-47s & L-5s – some of the types of planes glider pilots often wound up transitioning into when they were some place where there were no gliders! Ed won the DFC, air medal with cluster and the Purple Heart and left active service in July of 1945. Ed's well worn A-2 leather jacket can be found on permanent display at the Silent Wings Museum in Terrell, Texas.

Right: Front view of flight officer Ed Cook's A-2 jacket. Note fuselage marking on CG-4A glider in the background. (Silent Wings Museum)

Left: Left breast of Ed Cook's A-2 with impressed leather name tag and unusual 2-piece leather 14th Air Force patch. Design is apparently painted on the soft leather center and a stronger "chrome" leather rim is applied around the edge to make it more durable. This detailed and unusual production technique is typical of hand made items generated from natives and tailor shops in remote and distant areas of operation during the war. (Silent Wings Museum)

R.F. CRAWFORD

Lt. Crawford's A-2 has all leather insignia, including an American flag, C.B.I. patch, 10th Air Force patch, 332nd Troop Carrier Squadron (later the 11th Combat Cargo Squadron) insignia, rank, and name strip with wing.

Front of A-2 named to R.F. Crawford, 332nd Troop Carrier Squadron (later the 11th Combat Cargo Squadron), 513th Group, 10th Air Force. (JS Industries)

Rear of Crawford's A-2 with leather U.S. flag. (JS Industries)

C.C. DENAR

The only information given by C.C. Denar's A-2 is that he served as a pilot in the 5th Air Force. Denar was obviously proud of the 5th, as his jacket has that insignia on the chest and shoulder. His name tag has a silver pilot's wing over the name.

A-2 jacket of C.C. Denar, pilot, 5th Air Force. (JS Industries)

C.A. FARRAND

C.A. "Dolph" Farrand was a B-25 pilot with the famed "Burma Bridge Busters," 490th Bomb Squadron, 341st Bomb Group, Tenth Air Force. The 490th developed a technique called "hop" bombing, which made a major contribution to crippling Japanese lines of communication in Burma. The technique was highly effective at destroying bridges, which earned them their name. Unfortunately, Dolph's jacket is not around anymore, but we do have photos and the outstanding insignia which was once sewn on it. See also dust jacket front.

Leather C.B.I. patch from left shoulder of Farrand's A-2.

Leather Tenth Air Force patch worn on right shoulder of Farrand's A-2.

Multi-piece leather back patch from Farrand's A-2 jacket. The mission scoreboard indicates 44 missions with the "Burma Bridge Busters."

Multi-piece leather squadron patch of the 490th Bomb Squadron from Dolph Farrand's A-2.

Left: Sgt. Alan Nichols, flight engineer with the 490th Bomb Squadron wearing a very similar A-2 jacket to Dolph Farrand's. He is standing by the nose of a B-25J. (C.A. Farrand)

Right: The C.O. of the 490th had this B-25H stripped of guns and specially painted with the 490th skull and wings, for personal use. (C.A. Farrand)

China-Burma-India & the Pacific Theatre

W.A. LENTZ, JR.

W.A. Lentz was an American airman who served in the C.B.I. with the 14th Air Force. That is all we can determine from his A-2 jacket.

Left: A-2 jacket worn by Lt. W.A. Lentz, Jr., 14th Air Force. (JS Industries)

Below: Detail of "W.A. Lentz, Jr." painted over leather name tag and beautiful silk 14th Air Force patch on left chest. (JS Industries).

Left: Interior of Lentz's A-2 revealing a Chinese blood chit and American flag. (JS Industries)

J.P. McGOVERN

All the available information about 2nd Lt. J.P. McGovern is that he flew P-38 Lightning's in the C.B.I. theatre. His B-10 jacket has a very simple painting of a P-38 on the back.

Left: P-38 Lightning painted on the back of a B-10 jacket worn by 2nd Lt. J.P. McGovern.

C.B. MOORE

The 25th Bomb Squadron flew combat missions in the C.B.I. theatre and the western Pacific in B-29s. They were stationed in Chakulia, India and at West Field, Tinian. The 25th was assigned to the 40th Bomb Group, 20th Air Force. C.B. Moore was an American airman in the 25th Bomb Squadron. Moore's A-2 has all leather insignia on the outside, and a cloth blood chit inside.

Right: A-2 jacket worn by C.B. Moore, 25th Bomb Squadron, 40th Bomb Group, 20th Air Force. Patch on the left front panel is the 25th Bomb Squadron. Below: Inside with U.S. flag blood chit sewn in place. (both - JS Industries)

E.A. SIEGEL

The 12th Bomb Group "Earthquakers" was assigned to the 9th, 12th, and finally, the 10th Air Forces. They served in Egypt, Libya, Tunisia, Sicily, Italy, and India. The patch on Siegel's A-2 is that of the 83rd Bomb Squadron, which was part of the 12th Bomb Group throughout World War II. The patch has been changed, so it is likely that he was previously assigned to a different squadron. The 83rd flew B-25 Mitchells throughout the war and in 1945 received A-26 Invaders. E.A. Siegel's jacket has an amazing record painted on the back, of 63 bombing missions.

Right: A-2 jacket worn by E.A. Siegel, 83rd Bomb Squadron, 12th Bomb Group. Right below: Back of Siegel's A-2 with amazing 63 mission bomb record and 12th Bomb Group "Earthquakers" motif. Below: Leather 83rd Bomb Squadron patch and name strip on E.A. Siegel's jacket. (three - JS Industries)

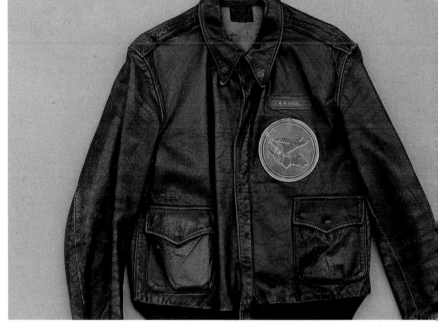

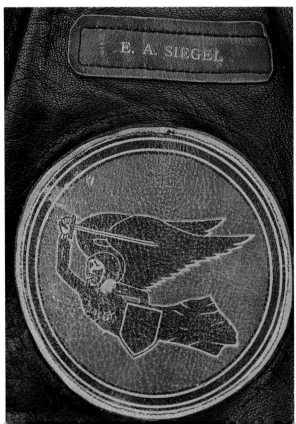

China-Burma-India & the Pacific Theatre

Left: Lt. William Neilson (top row, far left) with his crew.

Right: Front view of A-2 jacket worn by William Neilson, 491st Bomb Squadron, 341st Bomb Group, Tenth and 14th Air Forces.

WILLIAM NEILSON

Lt. William Neilson served as a B-25 pilot with the 491st Bomb Squadron, 341st Bomb Group, Tenth and 14th Air Forces. The 491st served in India and China and were primarily involved in operations to disrupt Japanese supplies and communications. After his service in the C.B.I., Neilson was assigned to Kirtland Field as a pilot for bombardier training. Bill Neilson's A-2 jacket is truly a classic example of a C.B.I. A-2. At one time, the "blood chits," or escape flags, were sewn on the back. As it was discovered these made good targets, the chits were moved inside the jacket and also served an additional function as pockets.

Below left: American flag "blood chit" sewn inside the left front panel of Neilson's A-2. The flag is left open at the top to serve as a pocket. Right: Chinese "blood chit" pocket sewn inside right front panel of Neilson's A-2.

Left: Multi-piece leather C.B.I. patch sewn to left sleeve of Neilson's A-2.

Right: Multi-piece leather A.A.F. patch sewn on right sleeve of Neilson's jacket.

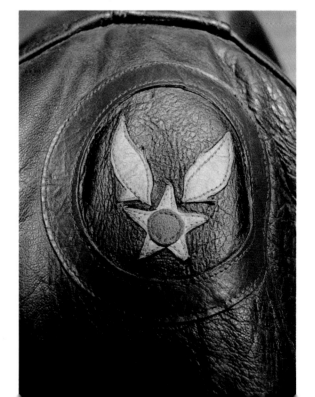

J.F. SHOAF

Right: The A-2 jacket worn by Lt. Shoaf in this photograph is unusual in that Shoaf has his name tag on the right chest instead of the customary left. The squadron insignia under his name tag is the 1st Bomb Squadron, which was assigned to the 9th Bomb Group. The 9th was a testing and training unit until early 1944 when they were equipped with B-29s and assigned to the 20th Air Force for combat in the Pacific Theatre. In the Pacific the 9th Bomb Group operated from North Field, Tinian. The patch on the left chest of Shoaf's jacket depicts "Uncle Sam" in a gunfighter stance, wearing a cowboy hat and standing on top of the world. It is the insignia of the 677th Bomb Squadron. (Campbell)

DAVID E. SMITH

David Smith completed flight training prior to the outbreak of WWII and served as a pilot with the 430th pursuit squadron stationed at Fairfax Field in downtown Kansas City, Missouri. The 430th was disbanded prior to the war and soon thereafter David was activated, with many others, into regular army service. Of pre-war vintage, David's A-2 jacket is not unlike most but it does have a commercial, rather than military issue label. This may be due to the fact that the 430th was a part of organized reserves rather than a regular army unit and was not supported by even government monitored supply sources. Apparently David had to purchase this jacket on his own initiative but the jacket itself conforms to issue standards in every detail. When examined closely, a trace of the 430th insignia can be seen painted underneath the 81st Bomb Squadron patch now sewn to the left chest. David also wore this A-2 throughout his flying days with the 81st Bomb Squadron while piloting B-25s in the CBI.

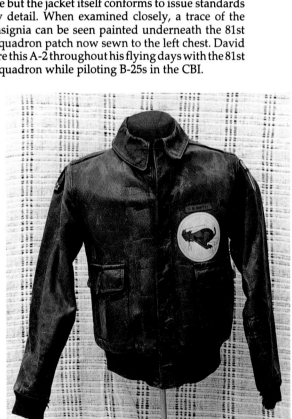

Front view of Lt. David E. Smith's A-2 jacket.

Rear view of Lt. David E. Smith's A-2 jacket.

China-Burma-India & the Pacific Theatre

...ose up of hand-made leather lieutenant's bar d Army Air Force patch on the right shoulder of iith's A-2 jacket. Hand-made leather insignia s a very common commodity from native rkshops in the CBI during the war and could en be bartered for with a package of cigarettes a stick of gum!

Close up of impressed leather name tag and hand-made leather 81st Bomb Squadron patch sewn to the left chest of Smith's A-2 jacket.

Close up of hand-made leather lieutenant's bar and China-Burma-India theatre patch on left shoulder of Smith's A-2 jacket. Endless varieties of native manufactured CBI patches, in all sorts of materials, can be found on service clothing of the second world war.

lose up of the 10" x 12" hand-made leather American flag sewn to the back of Smith's A-2 cket. The American flag was a very popular item of unofficial insignia on flying jackets during e war.

Original example of a 430th Pursuit Squadron (Organized Reserves of Missouri) patch. This patch is hand painted with the same design found under the leather 81st Bomb Squadron patch sewn to the left chest of Dave Smith's jacket. Unlike this example, Dave's was painted directly on the jacket.

R.N. TINELLO

Not much information is available on R.N. Tinello. His name and service number are stamped on the blood chit that came with this small group of material. He served as an American airman with the 375th Bomb Squadron, 308th Bomb Group, 14th Air Force, on a B-24 Liberator. The 375th flew out of China and India.

Leather multi-piece construction 375th Bomb Squadron patch from R.N. Tinello's jacket.

Leather C.B.I. patch from the shoulder of Tinello's jacket.

Blood chit named to R.N. Tinello.

An American airman of the 375th Bomb Squadron, 308th Bomb Group, 14th Air Force, wearing an A-2 jacket very similar to what Tinello's would look like with the insignia in place.

"Miss Beryl," a B-24 of the 375th Bomb Squadron.

Insignia of the 375th Bomb Squadron, 308th Bomb Group, 14th Air Force, painted on a B-24 Liberator.

China-Burma-India & the Pacific Theatre

Below left: A-2 jacket named to W.F. Whitley, Major, U.S.A.A.F. Right: Back of Whitley's A-2 with cloth U.S. flag and Chinese blood chit. (both - JS Industries)

W.F. WHITLEY
Major W.F. Whitley was an American airman who served in the C.B.I. At the time of this writing, his squadron insignia is unidentified, but his jacket is none-the-less spectacular with two blood chits on the back.

UNNAMED C.B.I
This A-2 jacket is a beautiful example of a C.B.I. jacket with blood chits sewn on the back. There is a nice variety of theatre made insignia on this piece.

Printed on cotton, theatre made C.B.I. patch sewn to left shoulder of unnamed A-2. (Arthur Hayes)

Front view of unnamed C.B.I. A-2 jacket. (Arthur Hayes)

Rear view of unnamed C.B.I. A-2 jacket. (Arthur Hayes)

UNNAMED 14TH AIR FORCE

This A-2 jacket is a typical example of a 14th Air Force piece. The unusual aspect of the jacket is the beautifully made bullion 14th Air Force patch on the left breast. The jacket also has leather multi-piece construction shoulder patches and two blood chit/map pockets added inside.

Unnamed 14th Air Force A-2 jacket. (Mike Conner)

Bullion, cloth, and leather 14th Air Force patch sewn on unnamed A-2. (Mike Conner)

Leather A.A.F. patch on left shoulder. (Mike Conner)

Leather C.B.I. patch on right shoulder. (Mike Conner)

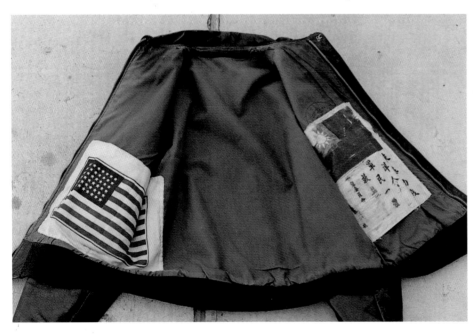

Inside of unnamed 14th Air Force A-2 showing placement of blood chit/map pockets. (Mike Conner)

Right: Detail of Chinese flag blood chit. (Mike Conner)

THE 11TH BOMB GROUP

The "Lady from Hades" and her crew, top row: Weaver (radar), Tom Huggins (engineer/top turret), Paul Laine (radio/waist), Fred Cornett (tail gunner), Claude Webb (ball turret), Joe Shay (waist); bottom row: James A. Perfetti (navigator), Burt Sheriff (co-pilot), Irving Freedenberg (bombardier), Jack Estes (pilot), Perry (nose gunner). (John Campbell)

The 11th Bomb Group was assigned to the 7th Air Force in 1943 and flew B-24 Liberators. The Group participated in the air offensive of the Gilberts, Marshalls, and Marianas while flying from Funafuti, Tarawa, and Kwajalein. In October of 1944, the Group moved to Guam, and in July 1945, to Okinawa.

BURT SHERIFF

Burt Sheriff flew his first combat mission November 12, 1944 with the 392nd Bomb Squadron, 30th Bomb Group, 7th Air Force, from Saipan. February 26, 1944, he transferred to the 26th Bomb Squadron, 11th Bomb Group, 7th Air Force. Burt completed his tour with 40 missions, as the bombs on his A-2 indicate. Burt's A-2 jacket provides a nice example of nose art being duplicated on a jacket. "Lady from Hades" was the name given to the crew's B-24 by the bombardier, Irving Freedenberg. Sheriff provides an outstanding record of his missions on "The Lady from Hades," which follows:

Oct. 12, 1944 - From Oahu to Midway Island. Flew out at 8500 ft., back at 9500 ft. Land on Sand Island which was larger of islands. Tower was Alcatraz. Flew out 7 hrs 45 mins. return on 13th [?] in 7 hrs. 35 min. Had trouble with gas leaks at take off on return.

392 Bomb Sqdn. 30th Bomb Group - Saipan
1. Nov. 12, 1944 - quote from my diary -
 "Our first mission was very interesting but not too bad. Flak was meager and as far as our ship was concerned inaccurate. Bill put our bombs right on target which was a specified portion of Iwo Jima. We were all pretty tired when we finally got back. Ship - "Thar she Blows II." Saw 3 enemy interceptors but they let us alone except for phosphorous bombs."
 8 hrs. 35 minutes

2. Nov. 22, 1944 - 7 hrs 55 min - Truk
 from the diary "This was our first mission in our new ship. She flys like a dream, at least as near like it as is possible for a B-2A. Mission over Truk with P-38's as escorts. Got 6 Zekes but P-38 got five of them. Webb claims the sixth but I don't think he will get it. Flak was meager. I couldnt see much because I flew over the target. Bombs on Ack Ack positions."

3. Nov. 28, 1944 - 8 hrs 10 min - Iwo - 3 hrs night
 "Did another good job on Iwo Jima causing fires and explosions. Two interceptors. One of them made a pass at us and flipped a bomb on us. Our boys let him know he wasnt very safe so he left. Flak was moderate and accurate. A few of our boys shot up."

4. Dec. 3, 1944 - 8 hrs 20 min - Iwo - 8 hrs 20 min night
 "Really a dilly in the line of a night snooper over Iwo Jima. Everthing from navigation to bombing done by radar. Bombed in over cast at 6,000. Our radar window threw off search lights. Bombs were Frags and saw them burst on target in break a way. Nice mission all alone. Thar she blows III.

5. Dec 5, 1944 - 2 hrs 50 min - Pagan
 "Milk run de lux. Whole trip about 3 hours. I flew left seat. Bill salvoed bombs on runway. 5000 lbs of T.N.T. Mission over Pagan where they think Japs may stage for raids here. No flak, no fighters."

6. Dec. 8, 1944 - 8 hrs 10 min - Iwo
 "Pearl Harbor Day back home so B-29s, P-8s- and 2 groups of B-24s pound Iwo from the air while cruisers hit from sea. Poor Japs. Bombed through undercast so didnt get results till cruisers report their planes couldnt watch fire because debris was going 800 feet high. No resistance."

7. Dec. 12, 1944 - 8 hrs 20 min - Iwo
 "I guess the japs on Iwo Jima are pretty unhappy with us. Today they really gave some opposition. Flak was moderate and accurate. Roughest fighters so far too. First time I have heard flak and I hope its the last. Nearly half our ships were shot up but not us. Good Bombing. This is a good one to talk about."

8. Dec. 17, 1944 - 9 hrs 20 mins - Iwo
 "Flew around for an hour and fifteen minutes at 19,500 trying to get radar to find island. Saw one fighter trying to throw out window to mess up radar. Bombs were dropped on island but thats all we know. It was not the kind of mission I like. Iwo Jima."

9. Dec. 24, 1944 - 10 hrs 5 min. Chi Chi Jima
 "Went to Chi Chi Jima this time expecting a rough mission but the island was socked in so we bombed radar and just got radar flak which was meager. We led B Flight. Results unobserved, no fighters. Sure was a long tiresome mission. carried 100 lb G.P. with Booby Fuses."

Dec. 27, 1944 - 3 hrs 15 min - abort to Iwo.
 "Started to Iwo but lost #3 a little past Pagan. Came back after salvoed bombs."

Burt Sheriff, 26th Bomb Squadron, 11th Bomb Group, 7th Air Force, in front of "Lady from Hades." (Sheriff)

Lt. Burt Sheriff, U.S.A.A.F. (Sheriff)

Left: Crew of the "Lady from Hades" under her wing. The tail marking is the 392nd Bomb Squadron, 30th Bomb Group, which is where they were assigned before transferring to the 26th Bomb Squadron, 11th Bomb Group. (Sheriff)

A-2 jacket worn by Burt Sheriff, 26th Bomb Squadron, 11th Bomb Group, 7th Air Force.

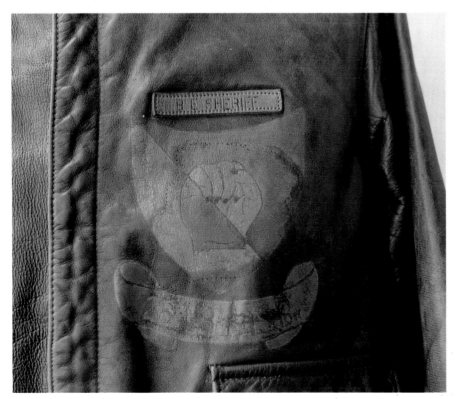

Detail of leather name strip and painted 26th Bomb Squadron insignia on left chest of Sheriff's A-2. Stitch holes remain from the 392nd Bomb Squadron patch, which was removed after he was transferred to the 26th.

*10. Jan. 3, 1945 - 9:00 hrs. Iwo - 2:25 night

"Went to Iwo for a late afternoon raid. Carried 40-100 pound G.P. Everything went swell till we got over target then we really got it. One fighter for Ack Ack which was intense and accurate-which is a lot of flack. Nine planes in our squadron with five damaged. We got our hydraulic system shot out and had to crank down gear, pump flaps down and sweat out the one accumulator for brakes. Cornett rigged up a chute in case the brakes didnt hold out. Everything worked out all right though."

11. Jan. 5, 1945. 8 hrs 35 min - all at night - Iwo

"This was a mission I dont care to talk about. We flew a snooper over Iwo and didnt have radar so had to drop on E.T.A. and I think we missed the island completely. We get credit for it but as long as we have to go that far I'd like to do some good."

12. Jan. 9, 1945 - 8 hrs 10 min. - all at night - Iwo

"This was the kind of mission I like to fly. We pulled a snooper over Iwo again and bombed radar. We saw our bombs hit right in the target area and didnt have trouble with anything. Every time I see the bombs hit the target I feel like the mission was pretty good.

We are sure getting tired of this tongue for chow."

13. Jan. 11, 1945 - 8 hrs 45 min - Iwo

"We have a rough mission with no Jap opposition today. Bombed Iwo with 100 pounders and flew the whole formation through cumulo nimbus clouds so thick we couldn't see more than about 40 feet. Our "A" flight was the only one that didnt get broken up and about 1/2 the time we couldnt see the lead ship. Cornett and Laine flew the airplane a while on the way home."

*14. Jan. 14, 1945 - 7 hrs 55 min - Truk

"We hit North Moen Field in the Truk Atoll. Had to bomb by radar but could see enough to see bombs on runway. Got jumped by 5 fighters but P-38 escort got 5 of them. One of the japs bailed out but the P-38s fight just like the japs so I doubt if he was alive when he hit the drink. One fighter blew up pretty close to us - saw another spin in."

*15. Jan. 18, 1945 - 9 hrs 45 min - Ha Ha Jima

"The natives on Okinura Town on Ha Ha got a few fires today when we dropped 9 - 350 lb incendiaries in their laps. Our bomb hits werent very good but we had a lot of very bad weather. Not much flak, moderate and not many planes shot up. Started several fires and saw the flak positions."

16. Jan. 21, 1945 - 8 hrs 10 min - all at night - Iwo

"Another snooper over Iwo. Carried 9 - 500 lb Frag. Clusters. The island was socked in tight but we saw two big explosions so bomb hits must have been good."

17. Jan. 27, 1945 - 8 hrs 10 min - all at night - Iwo

"Iwo got snooped again tonight we got away from the island before a couple of fighters could catch us. 10 - 500 lb. Frags in the dispersal area Bingo.

18. Feb. 1, 1945 - 8 hrs 5 min - Iwo - 5 hrs night

"Snooped again without radar. I think bombs hit the island but give no guarantee. Its a mission but not the kind we like to fly."

*19. Feb. 3, 1945 - 9 hrs 25 min - Iwo - 4 hrs 25 min night

"Day light mission over Iwo. We led C flight and had a sad mission. Bombed by radar and had to make two runs because lead ship radar was not good. Bill saw bombs hit through a hole in undercast. Got jumped by some pretty eager fighters. Made several shooting passes and tossed a lot of phosphorous bombs into formation. Not much flak and not many airplanes damaged."

20. Feb. 7, 1945 - 9 hrs - Iwo - 2:30 night   3:00 instruments

"What a way to celebrate my 21st birthday. We had the worst front yet with bad iceing conditions. At 16,000 ft we were pulling 45 HG, 2600 R.P.M. and couldnt hold our altitude. Our wing men didnt stay with us when we went in the weather so they couldnt find us to bomb. Bomb hits unobserved but we probably did some good."

21. Feb. 13, 1945 - 8 hrs 20 min. all at night - Iwo

"Had another snooper over Iwo with good results. Nothing of interest happened to us. Had a dilly of an explosion in the middle of the bomb pattern. I'll bet the Japs dont like us."

*22. Feb. 15, 1945 - 8 hrs 25 min. Iwo

"If I have any grandchildren I want to be sure and tell them this one. We had a daylight raid on Iwo and the target was undercast so we went down to spot the bombs. The hits were pretty good, but the 5 fighters that jumped us made an exciting mission out of it. I think we got one fighter. We picked up a twenty MM. dud in #3 about three feet from my head. I am sure glad the japs make mistakes. It sure makes a guy feel sick to watch 6 - 20 MM cannon head right into him shooting like mad.

ROUGH WAR
REST LEAVE

Burt Sheriff, far left with a group of cadets, all wearing new A-2 jackets. The rag tied on Burt's leg was to prevent items from falling out of the pocket, as there were no flaps or zippers on this type of flight coverall. The aircraft is a PT-1 (Sheriff)

China-Burma-India & the Pacific Theatre

nsfered to 26th Bomb Sqdn. 11th Bomb Group 2/26/45 came to Guam from rest
e – as I remember.

March 15, 1945 - 11 hrs 25 min - 3 hrs night - Chi Chi Jima
"Went to Chi Chi. sure had a funny feeling when we passed Iwo.
edenberg pulled an awful stunt by dropping his bombs accidentally about 2 miles
n shore. Such is life. Fly 12 1/2 hours, get shot at, and kill thousands of fish."

March 24, 1945 - 10 hrs 20 min - Marcus
"This is the first time we hit Marcus. We flew a wing and our flight leader
de a sorry run and dropped his bombs in the water. Our flight didnt get shot up but
re of the others got hit bad. Its a mission I guess 11 1/2 hours work to bomb fish."
ril 7th I checked out in Navy F-6-F hellcat-quite an airplane

April 28, 1945 - 7 hrs 30 min - Truk
"35 days since we flew a mission and we barely made this one. Lost #4
o on take off and had to fly another plane. We started our engines as the formation
sed over the field. Got to Truk and tagged on the formation in the bomb run. We well
t up Param Island with 30 - 100 lb G.P.s, some with 6 - 12 hour delay. I imagine there
e a lot of jap purple hearts on that. The flak was moderate and accurate beating up
e good boys."

April 30, 1945 - 11 hours - 2 hours on instruments - Marcus
"Marcus again but we really pounded the runway. We started a nice big
- more than likely a fuel dump. Flak meager and accurate with a few airplanes
naged."

May 2, 1945 - 11 hrs 40 min - 50 min on instruments - Marcus
"Today we hit Marcus with 4 - 500 lb G.P. with air burst fuses. They
loded right over the A.A. positions and imagine they did a lot of damage. B flight
ried frags and started a fire in the dispersal area. Moderate and accurate flak causing
te a bit of damage."

May 3, 1945 - 8 hrs 15 min - Truk
"Another field day for grey hair cultivation, we hit Eten Island and
blon hit us. We got 3 direct hits from 127 M.M.A.A. guns, and a few other besides.
d holes in every gas tank when we landed but lost 600-800 gals from #2 & #3. 69 big
es in the airplane. Threw away guns, flak suits and all heavy stuff. This is hard on
an."

May 11, 1945 - 8 hrs 15 min - 1:45 night - Truk
"Bombed Truk again and I was pretty worried after last one. Had
ellent bomb hits. Flak moderate and accurate. Naturally we got hit."

May 13, 1945 - 7 hrs 45 min - Truk
"Hit Truk again to bomb caves used as hangars on South Moen Airfield.
bombed them 100% bomb hits by all flights. We carried 5 - 1,000 lb SAPs and we
lly sapped em. Flak meager and inaccurate."

y 16, 1945 - 11 hrs 10 min - 2:50 night - Marcus
"And so my children we come to a sad story—Poor navigation on the part
12 navigators no mission after 12 hours work."

May 20, 1945 - 10 hrs 25 min - Marcus
"Bombed Marcus through a solid undercast and while flying in a bad
nt. There is a lot more to combat than getting shot at."

May 25, 1945 - 10 hrs 25 min - Marcus
"Bombed Marcus. Freedenberg in the hospital so we took some other guy
t liked to got us killed. Lead ships bombs had malfunction so we made a run of our
n after going through flak once. This bombardier made about a two minute bomb
and we really got shot up."

May 28, 1945 - 11 hrs 10 min - Marcus
"Freedenberg still in the hospital so we took Brobeil. Bombed assigned
rts of runways on Marcus. 4 flights had 100% Bomb hits. Ahem. Really a beautiful
ht to see - Flak moderate and accurate."

June 2, 1945 - 8 hrs 30 min - Truk
"Flew with 98th today and they really screwed up. Made 4 radar runs in
ig front trying to bomb N. Moen. What a way to make a living. Flak meager &
ccurate."

Jun 12, 1945 - 11 hrs 10 min - Marcus.
"Another one with the 98th only this one over Marcus. Bombed through
dercast. Flak meager and accurate. Hits unobserved."

June 21, 1945 - 10 hrs 30 min - Truk 1:25 nite
"Another mission over Eten. Flak was moderate and accurate. Several
ps got hit pretty bad and McClellan lost an engine so we brought him home at 145
P.H.
When we got back home the island was socked in and it was night. We
w over the field but the lights were out because one of our ships cracked up on it. We
ssed a cliff about 50 feet.
Finally flew up and landed at B-29 strip. Had 50 gals of gas on down wind
g. Two ships cracked up including Col. Morrow.

June 24, 1945 - 11 hrs 20 min - Marcus
"Snooper over Marcus and is that a long lonesome ride. Beat up runways
etty bad and our evasive action was too much for the flak—meager and innacurate.

June 26, 1945 - 10 hrs 35 min - 5:00 night - Marcus
"Snooper over Marcus. Had a beautiful bomb run and dropped four 500
unders right on A.A. position on southern tip. We had air burst fuses so I am pretty
re we did a lot of damage.
Our wing man blew up on take off and only 4 men got out. A lot of good
n died that night. Went to funeral that afternoon. What a feeling to go over a target
owing your wing men have just died—it is war."

June 27, 1945 - 8 hrs 15 min - Truk
"A mission with the 42nd over Truk. Made radar run on N. Moen. Really
nessed up mission."

July 3, 1945 - 10 hrs 40 min - 2:30 night - Marcus
"This is it. A sorry mission because of malfunction but we are through.
arcus Flak moderate and accurate.

THE END

A NEW LEASE ON LIFE

Our airplane hit by AA or fighters.

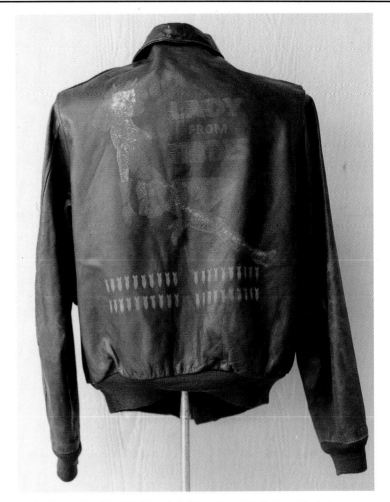

Back of Burt Sheriff's A-2. Note faded, but clear, "Lady from Hades" art work. Burt's 40 missions are represented by 40 yellow bombs.

7th Air Force insignia painted on left shoulder of Sheriff's A-2.

LONNIE J. EGGLESTON

Major Lonnie J. Eggleston, U.S.A.F. (retired), flew B-24s with the 98th Bomb Squadron, 11th Bomb Group, 7th Air Force. Eggleston flew 40 combat missions from 26, September 1944 until 8, March 1945. Strikes included Truk, Wake Island, Marcus, Iwo Jima, and Waleai. Major Eggleston also flew C-47, C-119, T 11, T-7, AT-6, and L-5 aircraft types. He ended his career with 4600 hours and combat time was 410 hours. Lonnie Eggleston's A-2 jacket has exceptionally high quality art work and it is in remarkably good condition. The back features a variation of a popular pin-up girl, which was painted on Guam. All insignia is painted, as well.

Lonnie J. Eggleston, as a Lieutenant at Cimarron Field, Oklahoma City, Oklahoma.

Painted insignia of the 11th Bomb Group, name, and pilot's wing on left chest of Eggleston's A-2.

Left: Front view of A-2 worn by Major Lonnie J. Eggleston, 98th Bomb Squadron, 11th Bomb Group, 7th Air Force.

Below: Detail of beautifully executed pin-up girl on the back of Eggleston's A-2. "My Baby" was also the name of his B-24.

Left: Back of Lonnie Eggleston's A-2 jacket.

China-Burma-India & the Pacific Theatre

THE 30TH BOMB GROUP – PRUETT & CHRISTOPHERSON

CAPTAIN EARNEST C. PRUETT

"Wee Willie" the bumble bee, carrying a red bomb, was the insignia of the 21st Bomb Squadron, 30th Bomb Group. This patch is seen on the left breast of Captain Earnest C. Pruett's A-2 jacket. Pruett flew B-24 Liberators with the 21st Bomb Squadron, 30th Bomb Group, Seventh Air Force.

Below: A-2 jacket worn by Captain Earnest C. Pruett, 21st Bomb Squadron, 30th Bomb Group, Seventh Air Force. (Michael J. Perry)

Squadron patch of the 21st Bomb Squadron, "Wee Willie," sewn to left breast of Pruett's A-2. The patch is unusual in that it is sewn to a piece of leather and then sewn to the jacket. (Michael J. Perry)

SGT. ARTHUR CHRISTOPHERSON

Sgt. Arthur Christopherson was a tailgunner on a B-24, "A-vailable," with the 27th Bomb Squadron, 30th Bomb Group, Seventh Air Force. His A-2 has a stylized rendition of his name and a hand painted caricature of a B-24 "dancing" in the clouds.

Detail of name and B-24 caricature painting on left breast of Christopherson's A-2. (Michael Perry)

A-2 flight jacket of Sgt. Arthur Christopherson, 27th Bomb Squadron, 30th Bomb Group, Seventh Air Force. (Michael J. Perry)

Crew of "A-Vailable" B-24. Arthur Christopherson is fourth from the left, front row, with circle around him.

90TH BOMB GROUP

"The Jolly Rogers," 90th Bomb Group, is one of the more famous bomb groups of the 5th Air Force. It was made up of the 319th, 320th, 321st, and 400th Bomb Squadrons flying B-24 Liberators. The 90th operated from Australia, New Guinea, Biak, the Philippines, and Ie Shima. The vertical stabilizers of their B-24's were marked with the skull and crossed bombs of "The Jolly Rogers."

N. STUART

N. Stuart was a bombardier with the 400th "Black Pirates" Bomb Squadron, 90th Bomb Group, 5th Air Force. His A-2 is complete with Australian made patches from each of the above units. Additionally, it has a leather name tag with bombardier wings.

A-2 jacket named to bombardier, N. Stuart, 400th Bomb Squadron, 90th Bomb Group, 5th Air Force. (JS Industries)

Below: 90th Bomb Group patch with black background for the 400th Squadron "Black Pirates." Below right: Leather bombardier's wing and name tag over "Black Pirates," 400th Squadron insignia on left chest of Stuart's A-2. (both - JS Industries)

China-Burma-India & the Pacific Theatre

Left: A-2 jacket from the 319th Bomb Squadron, 90th Bomb Group, 5th Air Force. (JS Industries)

UNNAMED 319th SQUADRON

This unnamed A-2 has a great example of the 319th "Asterperious" Squadron patch on the left chest and a nice 90th "Jolly Roger" on the right. The insignia is Australian made. Lt. Ken Strong had been an artist for Walt Disney and he drew the "Asty" character which became the squadron's insignia. The esoteric title of "Asterperious" came to mean "a superior attitude in an inferior environment."

Right: Six U.S.O. ladies wearing 90th Bomb Group A-2 jackets pose under the tail of a "Jolly Rogers" B-24 with the group marking clearly visible on the aircraft. The "Jolly Rogers" patch can be seen on five of the jackets and the 319th "Asterperious" patch shows up very well on the two jackets in the center. (John Campbell)

Australian made "Asterperious" 319th Bomb Squadron patch. (JS Industries)   *Australian made 90th Bomb Group "Jolly Roger" patch. (JS Industries)*

THE 380TH BOMB GROUP: "KING OF THE HEAVIES"

The 380th Bomb Group(H) was assigned to the 5th Air Force, but attached to the Royal Australian Air Force until early 1945. The 380th trained R.A.A.F. crews on B-24s. The 380th operated from bases in Australia, at first, and later moved to Mindoro and Okinawa. Some of the most spectacular flight jackets of the war are out of the 380th, as these examples prove.

WILLIAM H. HART

William Hart was an air gunner on a B-24 "I'll Be Seeing You" with the 529th Bomb Squadron, 380th Bomb Group, 5th Air Force. His jacket has a beautiful painting of a B-24 with the same nose art as the actual airplane. The group, squadron, and 5th Air Force insignia are also nicely painted on the jacket. There are 35 bombs falling from the B-24 representing 35 missions.

Right: A-2 jacket worn by air gunner William H. Hart, 529th Bomb Squadron, 380th Bomb Group. 5th Air Force.

Nose art on B-24 "I'll Be Seeing You" of the 529th Bomb Squadron, 380th Bomb Group. (Acord via Campbell)

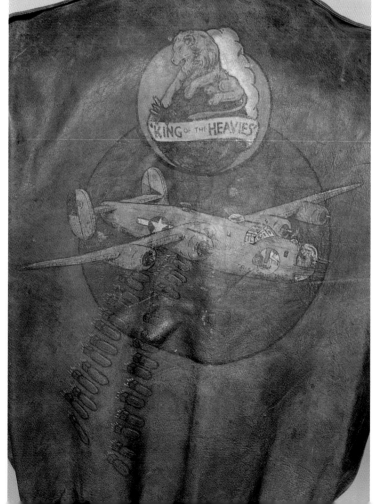

Right: Beautiful painting of 380th Bomb Group "King of the Heavies" insignia over B-24 "I'll Be Seeing You" dropping 35 bombs on the back of Hart's A-2.

China-Burma-India & the Pacific Theatre

GEORGE BECHSTEIN

George Bechstein served as an air gunner with the 529th Bomb Squadron, 380th Bomb Group, 5th Air Force. His jacket is unusual in that it is Australian made. Bechstein's A-2 is very similar to Hart's except there is no back painting on Bechstein's. Also, all insignia are sewn on Bechstein's jacket instead of being painted directly on the jacket.

Detail of leather 529th Squadron patch, air gunner's wing, and name tag sewn to left breast of Bechstein's A-2.

Australian made A-2 jacket worn by George Bechstein, air gunner, 529th Bomb Squadron, 380th Bomb Group, 5th Air Force.

UNNAMED B-10

This fantastic B-10 remains as a tribute to someone who flew 30 missions in a B-24 with the 380th Bomb Group. In this beautiful painting of the 380th Group insignia it is clearly visible that the lion is striking Japan with his paw as he sits on top of the world.

Left: B-10 jacket of the 380th Bomb Group, 5th Air Force. (JS Industries)

Right: Detail of 380th Bomb Group insignia painted on the back of B-10 jacket. (JS Industries)

SYL R. NEMCEK

Sylvester Nemcek was an air gunner with the 380th Bomb Group. The patch on the left chest of Nemcek's A-2 is the 531st Bomb Squadron, over which is painted his name and an air gunner's wing. The B-24 painted on the back of Nemcek's jacket is "I'll Be Seeing You" which, as stated earlier, was assigned to the 529th Bomb Squadron, 380th Bomb Group. It does not appear to have been painted by the same artist that did William Hart's jacket. It is likely that Nemcek transferred squadrons at some point in his career, which may explain this combination. Lack of contact with the original owner sometimes makes it difficult to determine exact circumstances behind unusual combinations of insignia.

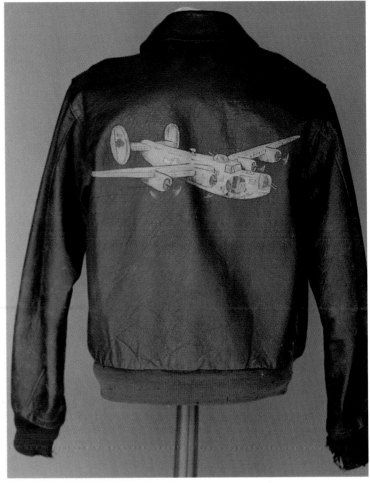

Back of Nemcek's A-2.

Left: A-2 jacket worn by Syl Nemcek, 380th Bomb Group, 5th Air Force.

Below: A-2 style jacket of the 435th Bomb Squadron. (Carmichael)

Below: Detail of embroidered on felt 435th squadron patch. (Carmichael)

435TH BOMB SQUADRON

The example from the 435th Bomb Squadron is an unusual private purchase A-2 style jacket. The 435th was part of the 19th Bomb Group early in the war and was stationed in Townsville, Australia. The Group came back to the States for most of the war and performed replacement training duties. The 435th Bomb Squadron later became part of the 333rd Bomb Group and was deployed in the Pacific too late for combat after training on B-29s. The only real claim to fame for the 435th was the evacuation of General MacArthur from the Philippines. The most likely origin of this jacket is that a B-29 crewman, who was never issued an A-2, decided he wanted one and had it made in Japan after the war. The squadron was stationed at Kadena and Okinawa, so this is a definite possibility. There is some speculation that the jacket was made in Australia, but it is not at all typical of an Australian A-2.

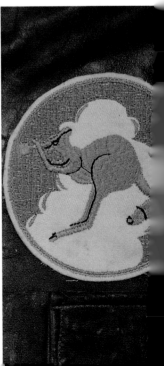

China-Burma-India & the Pacific Theatre

Detail of 5th Air Force patch and 8th Fighter Squadron patch, both likely Australian made, and faded leather wing on unnamed A-2. It is evident that at one time, a name strip was present.

UNNAMED 8TH FIGHTER SQUADRON

The 8th Fighter Squadron "Blacksheep," 49th Fighter Group, was assigned to the 5th Air Force. Like many fighter units, the 8th was stationed at a number of locations, including Australia, New Guinea, Biak, Leyte, Mindoro, Okinawa, and Japan. They flew numerous aircraft types, including P-35, P-40, P-47, P-38, and P-51. The name tag was removed from this jacket, so the original owner is now unknown. He was an American fighter pilot. The jacket is a clean example with two 5th Air Force Patches and an 8th Fighter Squadron patch. The leather strip above the patch has a faded wing on it.

Right: A-2 jacket worn by an American fighter pilot, 8th Fighter Squadron "Blacksheep," 49th Fighter Group, 5th Air Force.

THE 72ND FIGHTER SQUADRON

The 72nd Fighter Squadron was assigned to the 15th, 318th, and 21st Fighter Groups, all part of the VII Fighter Command "Sun Setters," 7th Air Force. The 72nd flew P-40s, P-39s, P-38s, and P-51s.

M.A. YATES

M.A. Yates was a pilot with the 72nd Fighter Squadron. His A-2 jacket has a leather name strip and a nicely painted 72nd Fighter Squadron insignia on the left front panel.

Left: A-2 jacket worn by M.A. Yates, 72nd Fighter Squadron, VII Fighter Command, 7th Air Force. (JS Industries)

V.J. SCHLOSSER

2nd Lt. Vernon J. Schlosser was a Mustang Pilot with the 72nd Fighter Squadron, VII Fighter Command, 7th Air Force. Schlosser is listed as killed in action 1 June 1945, while flying a P-51D.

VII "Sun Setters" patch as was once sewn to Schlosser's jacket.

A-2 jacket worn by V.J. Schlosser, 72nd Fighter Squadron, VII Fighter Command, 7th Air Force (Killed in action 1 June 1945). At one time the jacket had a VII Fighter Command patch on the right breast and a 72nd Fighter Squadron patch on the left.

Detail of stenciled Tokyo Club painting on the back of Schlosser's A-2.

Rear view of Schlosser's A-2.

Left shoulder of Schlosser's A-2 jacket with what remains of an Army Air Force decal.

China-Burma-India & the Pacific Theatre

LT. M.R. BECKER

Lt. M.R. Becker's A-2 jacket is unusual only in that it is an Australian issue A-2 jacket. Becker was most likely in the Fifth Air Force and probably acquired the jacket while on R & R in Australia to replace his original, which may have been lost or stolen. The Australian jackets were virtually identical to U.S. jackets with slight variations in lining material and color. Some Australian jackets were goat skin, but this one appears to be horse hide.

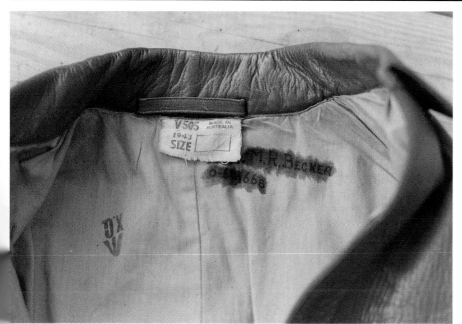

Label V505 1943 "made in Australia" in Becker's A-2 jacket. Becker's name and serial number are also visible, as well as the broad arrow proof mark. (Dale Edwards)

Right: Australian A-2 jacket worn by American Lt. M.R. Becker. The officers hat was not Lt. Becker's, but makes a nice companion piece, as it was made in Australia also. (Dale Edwards)

E.L. DONALDSON

Lt. E.L. Donaldson flew 37 missions as a navigator in B-29 Superfortresses with the 869th Bomb Squadron, 497th Bomb Group, 20th Air Force from Isley Field, Saipan. His A-2 jacket has a beautiful example of a chenille squadron patch and a faded 37 mission scoreboard painted under his leather name tag.

Detail of chenille 869th Bomb Squadron patch and leather name strip with faded 37 mission scoreboard painted below.

Right: A-2 jacket worn by Lt. E.L. Donaldson, 869th Bomb Squadron, 497th Bomb Group, 20th Air Force.

WILLIAM J. KERSCHER

Sgt. William Kerscher was an air gunner with the 8th Photo Recon Squadron, 6th Reconnaissance Group, 5th Air Force. The 8th flew many different aircraft, including P-38/F-4, P-38/F-5, B-17, and B-26. An interesting feature of Kerscher's jacket is that he recorded where he was stationed inside.

A-2 jacket worn by Sgt. William J. Kerscher, 8th Photo Recon Squadron, 6th Reconnaissance Group, 5th Air Force. (Michael J. Perry)

Detail of 8th Photo Recon Squadron patch on right front of Kerscher's jacket. (Michael J. Perry)

Interesting record of stations written in the lining of Kerscher's A-2 jacket – Wau-3-10-43, Port Moresby-4-12-43, Wewak-9-27-43. (Michael J. Perry)

Left: Unusual bullion gunner's wing on green cloth over felt 5th Air Force patch on Sgt. Kerscher's A-2. (Michael J. Perry)

China-Burma-India & the Pacific Theatre

JACK E. RECTOR

Jack Rector flew 40 missions as a B-24 waist gunner with the 23rd Bomb Squadron, 5th Bomb Group, 13th Air Force. The 23rd flew B-17s and B-24s from a number of island bases. Rector started his tour from Guadalcanal and ended it in the Philippines. Most of his missions were on a B-24 "Little Queen Mary," named for the pilot's daughter. Jack's A-2 jacket has all painted insignia, including squadron, group, air force, and back insignia.

A-2 jacket worn by Jack E. Rector, 23rd Bomb Squadron, 5th Bomb Group, 13th Air Force.

B-17E with the same 23rd Bomb Squadron insignia as is painted on Jack Rector's A-2. By the time Rector joined the unit, they were equipped with B-24 Liberators. (13th Air Depot)

Unofficial, but popular, variation of the 23rd Bomb Squadron insignia painted on right chest of Rector's A-2.

5th Bomb Group insignia and name scroll painted on left chest of Rector's A-2.

Back of Jack Rector's A-2.

Detail of B-24 and eagle superimposed on a shield painted on the back of Rector's A-2. The scroll, which is barely visible, reads "Bomber Barons."

Left shoulder of Rector's jacket with painted 13th Air Force insignia.

R.B. WILEY

R.B. Wiley served on a B-24 with the 372nd Bomb Squadron, 307th Bomb Group "Long Rangers", 13th Air Force ("Jungle Air Force").

Right: Detail of B-24 painting on Wiley's A-2 jacket. (JS Industries)

A-2 jacket worn by R.B. Wiley. The "Scottie" dog patch is the 372nd Bomb Squadron, and the red patch is an unofficial "Long Rangers" insignia. (JS Industries)

Reverse of Wiley's "Long Rangers" A-2. (JS Industries)

UNNAMED 29TH BOMB GROUP

The square "O" on the tail of the B-29 painted on this A-2 designates the 29th Bomb Group, 314th Bomb Wing, 20th Air Force. The 29th operated from North Field, Guam and was active in the air offensive of Japan. The original owner of this jacket flew 35 missions and was apparently credited with a Japanese fighter. The name "Bartlesville" (Oklahoma) is visible on the fuselage, which must have been his home town.

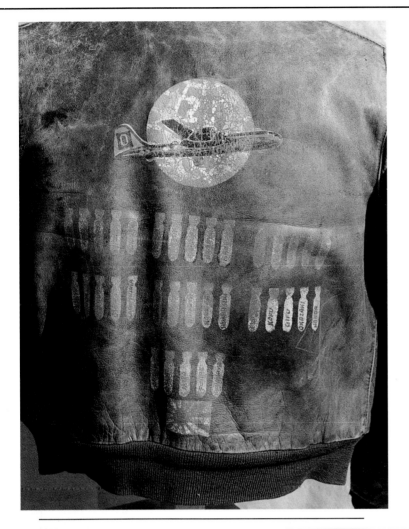

Left: A-2 jacket worn by a B-29 crewman, 29th Bomb Group, 314th Bomb Wing, 20th Air Force, likely from Bartlesville, Oklahoma. Note 29th Bomb Group B-29 superimposed on the 20th Air Force insignia. Also note the mission record of Japanese targets painted on each bomb. Also visible is the Rising Sun victory flag, representing a Japanese kill. (JS Industries)

GLIDER PILOT'S M-1943 FIELD JACKET

The Army quartermaster issue M-1943 field jacket was designed for wear by combat troops on the ground but many seem to have found their way into the air as an intermediate weight flying garment. This particular example was issued to a glider pilot who served in the Philippines as a member of the 1st Glider Provisional Group, 4th Section. Although the jacket itself is not unusual, the unique insignia of the 6th Section is hand painted on the back and makes it a rather exceptional example of the M-1943 jacket in a flying garment role.

Close up of the unofficial insignia of the 4th section insignia hand painted on the back of the Army issue M-1943 field jacket. It's interesting to note the tropical tan "Crush" type cap, flying scarf and glider pilot wings that have been incorporated into the "Bugs Bunny" design. Right: Rear view of the M-1943 field jacket worn by a glider pilot of the 1st provisional group-4th section in the Philippines. The drawstring closure is evident in the waist. (both - courtesy of Silent Wings Museum)

CHAPTER III

Europe, Africa, & the Mediterranean

The Air Forces engaged in Europe, Africa, and the Mediterranean were primarily the Eighth, Ninth, Twelfth, and Fifteenth. Generally speaking, the 8th and 15th Air Forces had the strategic mission of destroying and dislocating the German economy and war machine to a point where it could no longer function. The first mission flown by the 8th Air Force was August 17, 1942, when twelve B-17 Flying Fortresses hit Rouen. The Fifteenth Air Force was activated November 1, 1943 from units formerly assigned to XII Bomber Command. The first raid blown by the Fifteenth Air Force was November 2, 1943 to bomb aircraft factories at Wiener-Neustadt. The 8th and 15th formed a formidable team which played a major role in the destruction of German industry and in the defeat of the mighty Luftwaffe on the ground and in the air. These units were involved in air battles which became legendary – Ploesti, Schweinfurt, Regensburg, Berlin, Munster, Kiel, and Hamm, are but a few.

The heavy bomber units of the 8th and 15th flew B-17 Flying Fortresses and B-24 Liberators. The "little friends," as fighters were affectionately known, were primarily P-38 Lightnings, P-47 Thunderbolts, and P-51 Mustangs. The Mustang was, arguably, the most significant advance to the Allied air effort, as it was the first American fighter with enough range to accompany the bombers all the way to targets deep into enemy territory and back.

The 8th Air Force operated from bases in England and enjoyed relatively good living conditions. The Fifteenth flew from bases in Italy and many airmen lived in tents. Many men of the 8th had the luxury of time off in London, while veterans of the Fifteenth have fond memories of the Isle of Capri.

The Ninth and Twelfth Air Forces primarily had responsibility for tactical missions such as air-ground cooperation, bridges, sea ports, and operations designed to disrupt enemy communications. The Ninth and Twelfth also maintained responsibility for troop carrier and glider operations in the European Theatre of operations.

The weapons used by the Ninth and Twelfth included medium bombers and attack bombers. The B-26 Marauder, B-25 Mitchell, A-26 Invader, and the A-20 Havoc were used extensively. The Mustang, Thunderbolt, and Lightning were also used extensively in ground attack roles. Early in the war, the P-40 also saw a lot of action, especially in the North African Campaign.

Hand-painted 392nd Bomb Group insignia (see page 134). (JS Industries)

THE 94TH BOMB GROUP, 8TH AIR FORCE

The 94th Bomb Group was stationed at Bury St. Edmunds, England. Two fine examples of A-2 jackets from the 94th are "Texas Mauler" and "Skip." "Texas Mauler" was worn by T.M. Wright. The plain silver tail of the B-17 with black square A would indicate the jacket was probably painted around October/November 1944. Around January 1945 the group painted the entire vertical stabilizer yellow as seen on the back of "Skip." The 332nd Squadron patches on both jackets show a nice contrast in quality and style. Interestingly, the "Skip" jacket has a canvas patch, as seen on "Texas Mauler," sewn under the leather version of the same patch. "Skip" was S.W. Jenson's nickname. Jenson served as a bombardier with the 331st, 332nd, and 410th Squadrons. The markings on the B-17 on the back of Jenson's jacket are that of the 331st Squadron – dark blue cowlings from late 1944.

The troop carrier plane most heavily used was the C-47 Skytrain. C-47's carried out transport, paratroop, glider tow, combat cargo, and medical functions. The gliders used in these areas most commonly were the Waco CG-4 Hadrian and the Waco CG-13.

The flight jackets worn by airmen in these areas included all types of art and insignia. Probably the most easily recognized jackets are those done in Italy, and especially the Isle of Capri. Characteristically, these designs included tooled leather, full color patches. Leather American flags on the shoulder are also very common on jackets from Italy. High quality paintings on the backs of flight jackets were done by street artists and shop vendors on the Isle of Capri. Designs on jackets from England run a full range, from cartoon characters to pin-up girls. English made patches were usually embroidered on felt. The quality of paintings, as in all theatres, depended on the talent of the artist you happened to run into!

Left: Front view "Texas Mauler" A-2. Note T.M. Wright name strip and painted canvas 332nd Bomb Squadron insignia on left front panel. 30 mission scoreboard is painted on right front panel. (Hayes)

Right: Rear view "Texas Mauler" A-2. Note detail of "Texas Mauler" with B-17 depicted over map of Europe. (Hayes)

Left: Overall front view of "Skip" A-2. A nice touch on this jacket is the drilled dice attached to the zipper. Note left front panel of "Skip" with S.W. Jenson name strip and painted leather patch of the 332nd Bomb Squadron. The leather patch is sewn over a canvas version, as seen on "Texas Mauler."

Right: Overall rear view of "Skip" A-2. Aircraft markings indicate a B-17 of the 331st Bomb Squadron, 94th Bomb Group from late 1944.

Europe, Africa & the Mediterranean

94TH BOMB GROUP "B-17"

This A-2 jacket was purchased at a clothesline sale for a quarter in the early 1970s. Its original leather name tag has long since been removed and its cuffs are missing but its simple art work remains a tribute to someone who did 35 missions over "Fortress Europe." Although it bears a rather boring rendition of an outstanding aircraft, the artist did include the 94th Bomb Group's marking, square "A", on the stabilizer. This type of clue is often the only link that can ever be established with the original owner. This group had the distinction of being under the command of then Colonel Frederick W. Castle, 1 of 17 medal of honor winners in the 8th Air Force, he was awarded posthumously for an action while a Brigadier General acting as an Air Task Force Commander of the 4th Combat Wing on December 24, 1944.

Close up showing the simple rendition of a B-17 in flight over 35 mission tally "bombs." All artwork has been executed in a bright silver-metal color paint. Note square-A marking of the 94th Bomb Group on the stabilizer and nationality star on bottom side of starboard wing.

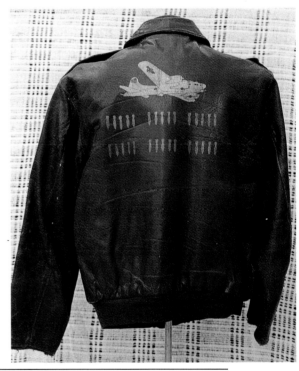

Right: Rear view of the 94th Bomb Group marked A-2 jacket.

96TH BOMB GROUP, 8TH AIR FORCE

FRED HUSTON

Fred Huston served as a bombardier on B-17 G 4-6170, "Sittin' Pretty," 337th Bomb Squadron, 96th Bomb Group, 8th Air Force. The 96th Bomb Group flew from Snetterton Heath, England. In regard to his A-2 jacket, Fred said, "It was issued to me in the Spring of 1944 at Kearney, Nebraska, which is as unlikely a port of embarkation as one could imagine, since the nearest thing to water in the vicinity is the Muddy Platte River. At Kearney we turned in the jacket that had been issued, in my case, at Advanced in Midland, Texas, and drew a nice stiff horsehide jacket to replace one that was already broken in."

Far left: Front view of Fred Huston's A-2 jacket. Left: Rear view of Fred Huston's A-2 jacket. (both - Jeff Huston)

Along with this photo came the following priceless note from Fred Huston: "Enclosed is something every young man should have: A photo of the airplane that won the war. This is Douglas B-17 G 4-6170, 337th BS, 96th BG (H). M/AW (Paintbrush M for Mike) starring in a strike photo, provided by my navigator, Rex Malut. I felt you should have a shot of the A/C that carried crew 3494 to fame and glory, led, of course, by the bombardier, Fred Huston, closely followed by the rest of the Crew. When Hitler and his evil craven aids heard that this airplane and crew were on the prowl over the Third Reich, they hastily beat a retreat to their concrete bunkers and began to have serious doubts as to the outcome of the war." (Fred Huston)

Fred Huston in the nose of a B-17 F flying out of Dyersburg, Tennessee during phase training. "The mask is the A-10, which was guaranteed to take the skin off the bridge of your nose in no less than ten minutes." (Fred Huston)

Fred Huston, bombardier, 337th Bomb Squadron, 96th Bomb Group (H), 8th Air Force.

Detail of "Sittin' Pretty" 35 times. "Sittin' Pretty" was a B-17 G. 35 times is the mission scoreboard. (Jeff Huston)

Nickname "Sooner" painted on left front panel of Fred Huston's A-2. "Sooner" comes from the University of Oklahoma. (Jeff Huston)

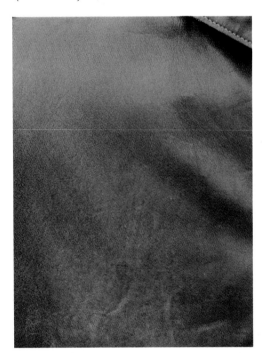

Right front panel of Fred Huston's A-2. Covered wagon is "The Sooner Schooner" – symbol of the University of Oklahoma, where Fred attended. The beer mug is also a reference to Fred's college days. (Jeff Huston)

Europe, Africa & the Mediterranean

R.A. BOHYER
S/Sgt. Bohyer was the tail gunner on the original crew of B-17 G "5 Grand" (the 5000th B-17 built by Boeing), 96th Bomb Group, 8th Air Force. Bohyer flew 35 combat missions as shown by the bombs on the back of his jacket.

Original crew of "5 Grand" under the nose art on the airplane.

Grouping of personal memorabilia from S/Sgt. R.A. Bohyer.

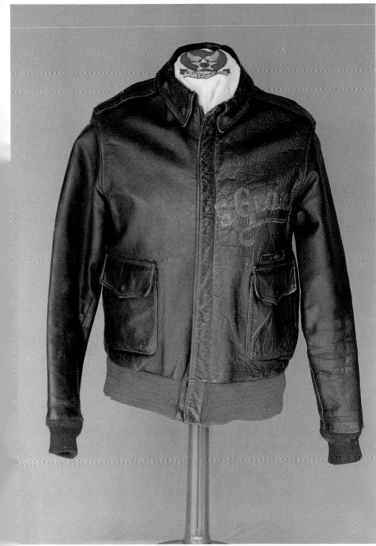

"5 Grand" A-2 worn by S/Sgt. R.A. Bohyer, 96th Bomb Group, 8th Air Force.

Back of "5 Grand" with B-17 G in flight and 35 mission bomb scoreboard painted on.

THE 86TH BOMB SQUADRON, 12TH AIR FORCE

The air gunner who wore this jacket apparently flew 38 missions of a 50 mission tour (it is difficult to tell exactly what he was representing with his bomb scoreboard, as there are 50 bombs, but only 38 numbered). The jacket was acquired from a family member so all details are not available. The dogs on the flaps represent the squadron mascot, and the map of Italy indicates where he served. The wing on the chest signifies air gunner and the squadron patch is the 86th Bomb Squadron, 47th Bomb Group, 12th Air Force. The paintings on the back of the jacket indicate he flew missions in A-20 havocs and A-26 Invaders. The A-20 was nicknamed "Miss Burma." The 86th Bomb Squadron flew out of numerous bases including Algeria, Tunisia, French Morocco, Sicily, Italy and France.

Right: Detail of painted A-20 Havoc, "Miss Burma", and A-26 Invader on the back of 86th Bomb Squadron air gunner's A-2 jacket. (JS Industries)

A-2 jacket worn by an American air gunner in the 86th Bomb Squadron, 47th Bomb Group, 12th Air Force. (JS Industries)

Reverse of 86th Bomb Squadron A-2. (JS Industries)

Europe, Africa & the Mediterranean

THE 100TH BOMB GROUP: "CENTURY BOMBERS"

The 100th Bomb Group is among the most well-known bomb groups of the 8th Air Force, having played a significant role in the Allied campaign against German aircraft factories during BIG WEEK 20-25, February 1944. The 100th also was on the famed Berlin Missions in March of 1944. The squadrons assigned to the 100th were the 349th, 350th, 351st, and 418th. The 100th Bomb Group flew B-17's from Thorpe Abbotts, England.

"PETE"

Pete was a B-17 pilot of the 100th Bomb Group with 33 missions. His B-17 was "Goin My Way."

Left: Detail of nicely painted pilot's wing and leather "Century Bombers" patch on left chest of Pete's jacket. (JS Industries)

Fantastic painting of a rabbit reclining on a bomb "Goin My Way" on Pete's A-2 jacket of the 100th Bomb Group. 33 missions are indicated on the tail of the bomb. (JS Industries)

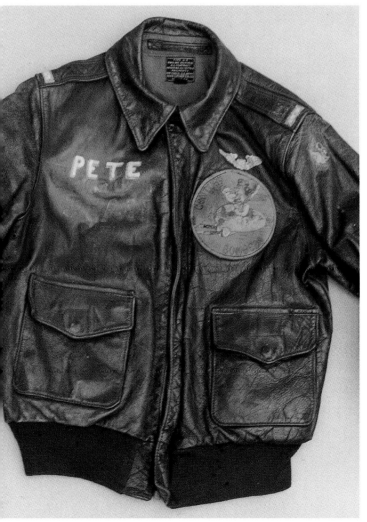

A-2 jacket worn by a B-17 pilot named Pete, 100th Bomb Group, 8th Air Force. (JS Industries)

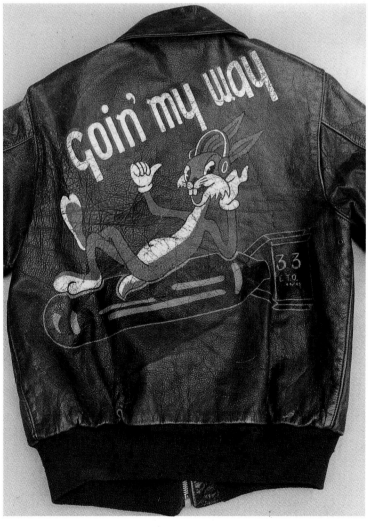

"DAVE"

All the information we know for sure about Dave is that he flew 35 missions in a B-17 with the 100th Bomb Group. His A-2 carries the name and painting of "Woody," which was likely the name of the airplane on which he served.

A-2 worn by "Dave." 100th Bomb Group, 8th Air Force. (JS Industries)

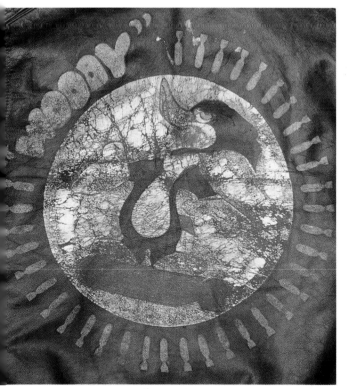

Detail of the back of Dave's A-2 featuring a painting of "Woody" standing on a bomb surrounded by 35 bombs indicating 35 combat missions. (JS Industries)

Back of Dave's A-2 jacket. (JS Industries)

Europe, Africa & the Mediterranean

Left: Front view of A-2 jacket worn by Kenneth Blackshaw, 423rd Bomb Squadron, 306th Bomb Group, 8th Air Force.

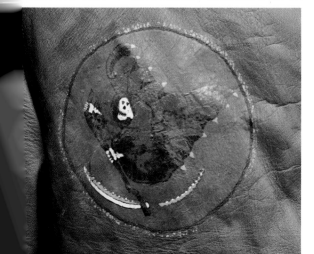

Left: Back of "The Gray Lady" with painted B-17 G in the markings of the 423rd Bomb Squadron, 306th Bomb Group. The number 35 in the swastika signifies 35 missions over Nazi territory.

Right: Crew of "The Gray Lady", serial number 43-38376: Front row; left to right, Ken Blackshaw, Wesley Gunkel, John Wilson, David Gorrell. Rear; Roy Nokes, Ed Tutun, Duane Brunner, Robert Maphis, Forrest Yorgason, Charles Yeager.

Right: Left breast with leather name tag "K.D. Blackshaw" affixed.

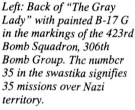

Left: Right breast of "The Gray Lady" with painted "Firey Phantom" insignia of the 423rd Bomb Squadron.

THE 306TH BOMB GROUP, 8TH AIR FORCE

"THE GRAY LADY"

Nantucket Island was known as "The Gray Lady." It was also the home of Ken Blackshaw, who gave that name to a B-17 G in the 423rd Bomb Squadron, 306th Bomb Group, 1st Air Division, Thurleigh, England. The serial number of "The Gray Lady" was 4338376.

Lieutenant Colonel Kenneth D. Blackshaw, USAF (Retired), was the first pilot of "The Gray Lady" and he painted the jacket himself.

Ken Blackshaw graduated from flight training in the class of 44B at Turner Army Air Field, Albany, Georgia. After graduating, he served as an instructor. At the time, he was 25 years old – an old man for a World war II pilot!

While assigned to the 306th Bomb Group, Blackshaw flew 35 missions, thus the 35 in the top of the swastika on the back of the jacket. This was his way of signifying 35 missions over Nazi territory. "The Gray Lady" was lead navigator ship for the 8th Air Force on two missions. On Blackshaw's first mission with "The Gray Lady," she "got nicked." The crew counted over 75 holes in the ship! The only purple heart on the crew was given to the original tail gunner. On that mission only one shell hit the airplane and also an artery in his arm. After hitting the target, "The Gray Lady" left formation and headed full throttle for England while the radio operator administrated first aid.

Ken Blackshaw stayed in the service for 20 years. Following the war he flew over 1500 hours in C-47's and ended up in KC 97's at Barksdale Air Force Base.

"The Gray Lady" jacket is a size 42 U.S. Army Air Force A-2 manufactured by Rough Wear Clothing Company, Middletown, Pennsylvania.

DUANE BRUNNER

Staff Sgt. Duane Brunner was the ball turret gunner on "The Gray Lady." Although Brunner and Blackshaw were on the same airplane, the insignia and design of the art work on their jackets is quite different.

Right: Detail of leather name strip and embroidered on felt "Firey Phantoms" 423rd Bomb Squadron patch on left chest of Brunner's jacket.

A-2 jacket worn by S/Sgt. Duane Brunner, 423rd Bomb Squadron, 306th Bomb Group, 8th Air Force.

Back of Brunner's "Gray Lady" jacket, which features the airplane superimposed on the 8th Air Force insignia with a flak burst and a bomb with the number 30 on it.

CHESTER W. LANTZ – "WE PROMISED"

Staff Sergeant Chet Lantz entered the Army in Houston, Texas and went overseas as an air gunner on B-17s of the 306th Bomb Group, 367th Bomb Squadron – stationed at Thurleigh, England with the mighty 8th Air Force. Chet was decorated for "Meritorious Achievement" in air combat, completed his combat tour and returned home without a scratch. Given the odds of survival as a waist gunner, this was quite a feat. Although the artwork is relatively simple, Chet's A-2 jacket is an interesting example. Of special interest is the similarity of the lettering on the jacket to that which can be seen on the aircraft. A good indication that one artist executed both jobs! The 367th Bomb Squadron patch on the left chest is most unusual as the monkey head motif was embossed in the leather prior to painting, not simply painted on a flat piece of stock.

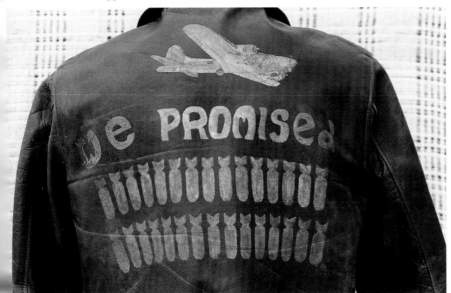

Close up of the Lantz A-2 breast insignia. The 8th Air Force design is hand painted directly on the jacket. The red-toned "loops" in the lower left wing were to indicate Lantz was a port side waist gunner. The monkey head insignia is embossed in the leather before painting giving it a two-dimensional appearance.

Front view of Staff Sgt. Chet Lantz's A-2 jacket.

Close up of B-17 in flight over nickname "We Promised" over 30 hard earned mission tally "bombs" painted directly on the leather in off-white enamel.

Rear view of Staff Sgt. Chet Lantz A-2 jacket.

Staff Sgt. Lantz "at work" in the waist of "We Promised."

"LADY LORRIE"

The fourth jacket from the 306th Bomb Group is also of the 423rd "Firey Phantoms" Bomb Squadron. Its original owner is now unknown, but he flew 35 missions on a B-17 "Lady Lorrie." The back painting has a female figure riding a B-17 in the markings of the 423rd Bomb Squadron, 306th Bomb Group, dropping 35 bombs. The blue tip on the tail, indicating the 423rd Squadron, dates the painting after August of 1944. The squadron patch on this jacket is embroidered on felt, while the 8th Air Force insignia is painted.

A-2 jacket "Lady Lorrie," 423rd Bomb Squadron, 306th Bomb Group, 8th Air Force. (Manion's)

Detail of painted 8th Air Force insignia and embroidered on felt, 423rd "Firey Phantoms" Bomb Squadron patch on "Lady Lorrie" A-2. (Manion's)

Back of 306th Bomb Group A-2 "Lady Lorrie." (Manion's)

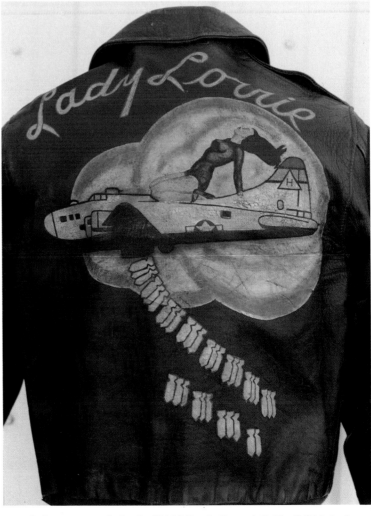

Detail of painting on the back of 306th Bomb Group A-2. The B-17 is in the markings of the 423rd Bomb Squadron, 306th Bomb Group, and the bombs indicate 35 missions. (Manion's)

Europe, Africa & the Mediterranean

THE 351ST BOMB GROUP, 8TH AIR FORCE

The 351st Bomb Group, 8th Air Force, flew B-17 Flying Fortresses out of Polebrook, England. We were lucky to find six examples of A-2 jackets from this unit!

JOHN S.L. SWARTS – "PUGNACIOUS BALL"

The 511th Bomb Squadron (of the 351st Bomb Group) had a well-known tradition of naming their aircraft after the Squadron Commander, Major Clinton F. Ball. This resulted in a naming series for the units fortresses – Cannon Ball, Screwball, Highball, Linda Ball, Pugnacious Ball, etc. Staff Sergeant "Johnny" Swarts was the port side waist gunner on the Pugnacious Ball for 6 missions. Swarts stayed behind on the ship's 7th mission and it was lost with all aboard over Achenshaven, Germany. Swarts was personally interviewed by the King and Queen of England while stationed at Polebrook, England and fondly recalls their sincerity and interest in the American airmen stationed there. The direct painted artwork on Swarts' jacket suggests the artist had some formal training.

Front view of Staff Sgt. John Swarts' A-2 jacket. One of his dog tags can be seen hanging from the right shoulder.

Rear view of Staff Sgt. John Swarts' A-2 jacket.

Close up of the attractive hand lettered style artwork painted on the back of Swarts' A-2. Note the shadow toning effect and the "511" that has been all but completely covered by the aircraft wing star design. "Ball Boys" was the group nickname for airmen of the 511th Squadron.

Close up of the left shoulder of Sgt. John Swarts' A-2 painted with the triangle-J marking used on the stabilizers of all 351st Bomb Group aircraft.

Original portrait photo taken of Staff Sgt. Swarts wearing his A-2 jacket and flying helmet. Nickname "Johnny" is barely visible on pocket in lower left hand corner.

"LORD'S ANGELS"

Not only was the name "Ball" popular in the 511th, apparently "Angel" was popular as well. The first example, "Lord's Angels," we have very little information on. This war time photograph of an American airman with 35 missions and the Triangle J of the 351st wearing his A-2 serves as the only known record of this jacket.

Unknown American airman wearing his A-2 "Lord's Angels." The jacket features a B-17 in flight, the Triangle J of the 351st Bomb Group, and 35 mission bomb scoreboard. (John Campbell)

A-2 jacket "Martha's Angel," 511th Bomb Squadron, 351st Bomb Group, 8th Air Force.

Back of A-2 jacket – "Martha's Angel."

Detail of 351st Bomb Group insignia on right chest and 511th "Ball" Squadron insignia on left chest of "Martha's Angel."

"MARTHA'S ANGEL"

The second aircraft to use the name "Angel" is "Martha's Angel." The lettering style and B-17 painting appear to have been done by the same artist that painted "Lord's Angels." Interestingly, the B-17 painting on "Martha's Angel" was done on fabric and sewn on as opposed to being painted directly on the jacket. The airman who wore "Martha's Angel" is also unknown, but he survived 35 combat missions.

Right: 8th Air Force patch painted on left shoulder of "Martha's Angel."

Far right: Detail of "Martha's Angel." Of special interest is the B-17 which is painted on fabric and sewn on the jacket. The Triangle J was the 351st Bomb Group marking on the vertical stabilizer of each aircraft. 35 bombs indicate as many missions.

Europe, Africa & the Mediterranean

A.J. SMETANA

Staff Sergeant Adolph J. Smetana was the tail gunner on a B-17 "Bigas Bird" with the 511th Bomb Squadron, 351st Bomb Group. Smetana flew 35 missions from 17 September 1944 until 20 January 1945, and was credited with one probable kill. A.J.'s jacket features the name "Bigas Bird," a B-17 in flight, an ostrich, and 35 bombs. Smetana painted the jacket himself.

Cloth 351st Bomb Group patch on right chest of Smetana's jacket. (Tim Smetana)

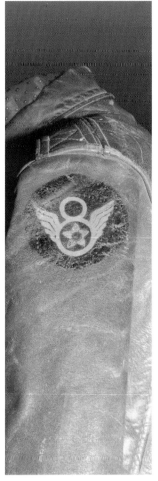

Short-winged version of the 8th Air Force insignia painted on the shoulder of Smetana's A-2. (Tim Smetana)

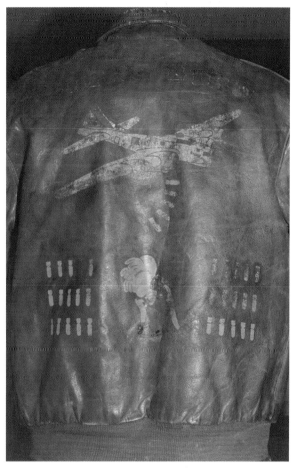

A-2 jacket worn by Staff Sergeant A.J. Smetana, 511th Bomb Squadron, 351st Bomb Group, 8th Air Force. Faded, but visible, are the words "Bigas Bird," a B-17 G, an ostrich, and 35 bombs. (Tim Smetana)

"WILDFIRE"

The airman wearing this A-2 flew 30 missions with the 351st Bomb Group on a B-17 named "Wildfire."

Left: Unknown airman of the 351st Bomb Group wearing his A-2 "Wildfire." The Triangle J identifies the 351st and the design is very much in keeping with other jackets from this unit. (John Campbell)

EDDIE HUCKLE

Eddie Huckle was a waist gunner on a B-17 "Pecks Bad Boys" with the 351st Bomb Group. The bombs on his A-2 indicate 25 missions. The painting on the back of his jacket is typical of the 351st, incorporating old English lettering, the Triangle J, and the U.S. National insignia.

Inside of Huckle's A-2 with name clearly visible. (Mitchell via Hatchel)

Detail of Huckle's jacket with "Eddie" painted in green over squadron patch on left chest and stylized 8th Air Force patch on left shoulder. (Mitchell via Hatchel)

A-2 jacket worn by Eddie Huckle, 351st Bomb Group, 8th Air Force. (Mitchell via Hatchel)

Back of Eddie Huckle's A-2. (Mitchell via Hatchel)

381st BOMB GROUP, 8TH AIR FORCE

WALLACE D. BECKMAN – "TOUCH THE BUTTON NELL"

Wallace D. Beckman was an air gunner with the 381st Bomb Group of the 8th Air Force in the E.T.O. during WWII. His A-2 jacket sports the name "Touch the Button Nell I and II" which strongly suggests he was a crewman on both planes of the same name. The 30 mission bombs painted on the back indicate he completed one combat tour with the 381st Bomb Group but little else is known due to lack of contact with the original owner. To further the mystery, the Running Devil Squadron patch was obviously worn but never authorized, making it difficult to determine which of the 4 squadrons of the 381st Bomb Group it may be. Establishing this jacket's affiliation with the 381st was only made possible by locating an original photo of the B-17 "Touch the Button Nell II" in the National Air and Space Museum archives which was marked to indicate it was in fact a ship from this group. Sometimes even the best detective work reveals far less information than this!

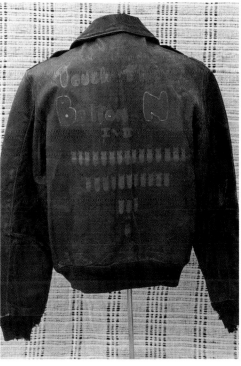

Front view of W.D. Beckman's "Touch the Button Nell I and II" A-2 jacket.

Rear view of W.D. Beckman's "Touch the Button Nell I and II" A-2 jacket.

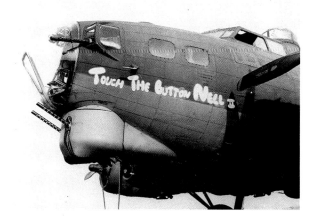

Period photograph of the B-17 "Touch the Button Nell II" identified to have been attached to the 381st Bomb Group. (NASM via Valant)

Right: Detail of the beautiful "letter jacket" Chenile Squadron patch on the left chest of W.D. Beckman's A-2 jacket. This style of embroidery resulted in a well wearing and colorful insignia of a type quite common during the second war.

Detail of the back of W.D. Beckman's A-2 jacket with aircraft name "Touch the Button Nell I and II" hand painted in glossy red paint. Note also the unusual configuration of the 30 mission bombs. It's interesting that the pigment has failed on only the paint used on the bombs. Often the bombs were not actually painted until the veteran had arrived home from overseas.

"TOMMY" – "ME AND MY GAL"

The second 381st Bomb Group jacket is named only to "Tommy." The jacket tells us that "Tommy" flew 35 missions on a B-17 named "Me and My Gal" and his girl's name was "Betty." The red wing tips, tail plane, and vertical band on the B-17 date the painting after July 1944.

Below: Chenille unit insignia on 381st Bomb Group A-2. (JS Industries)

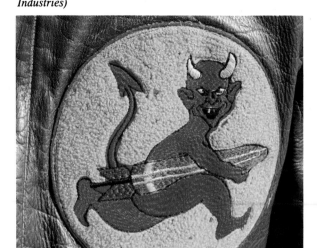

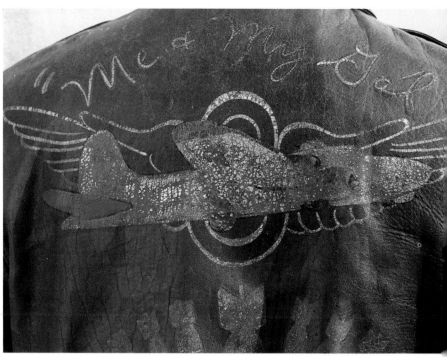

Above: Detail of B-17 G in 381st Bomb Group markings, 8th Air Force insignia, and "Me and My Gal" painted on Tommy's A-2. (JS Industries)

A-2 jacket worn by "Tommy" of the 381st Bomb Group, 8th Air Force. (JS Industries)

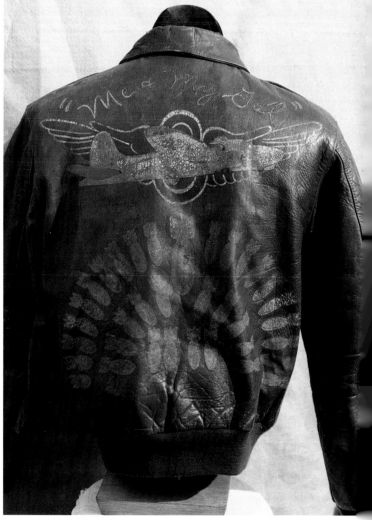

Back of 381st Bomb Group A-2 with "Me and My Gal" B-17 flying across the 8th Air Force insignia, over bombs indicating 35 missions. (JS Industries)

Europe, Africa & the Mediterranean

THE 392ND BOMB GROUP

ELDRIDGE "EDGE" THOMAS

T/Sgt. "Edge" Thomas flew 34 missions on a B-24 Liberator "On The Make" with the 576th Bomb Squadron, 392nd Bomb Group, 8th Air Force. He was the flight engineer/top turret gunner. The 392nd Bomb Group was stationed at Wendling, England.

Front view of T/Sgt. Thomas', 392nd Bomb Group, 8th Air Force, A-2 jacket. (Scott Thomas)

Rear view of Thomas' A-2 "On The Make." The painting on Edge Thomas' A-2 is of a B-24 flying through flak with 34 mission scoreboard and "On The Make." There is no bar or plus sign on the tail of the aircraft, which indicates the 576th squadron. (Scott Thomas)

T/Sgt. "Edge" Thomas, second from left, wearing his A-2 jacket in front of a B-24. (Scott Thomas)

Below: 392nd Bomb Group insignia, painted on leather, sewn to left breast of Thomas' A-2. (Scott Thomas)

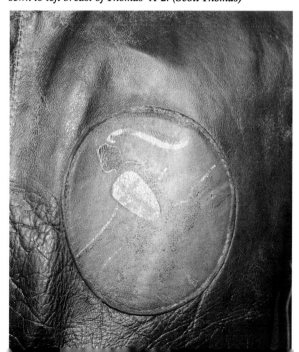

"HEAVENLY BODY"

The second jacket from the 392nd is named only to "Dick" from Long Island, New York, and his girl with a "Heavenly Body" was "Madeleine." "Heavenly Body" was also, most likely, a B-24 of the 392nd.

Right: Detail of high quality lettering "Heavenly Body" painted on the back of 392nd Bomb Group A-2 jacket. Note the name "Madeleine" painted in the star – no doubt, the girl with the "Heavenly Body." (JS Industries)

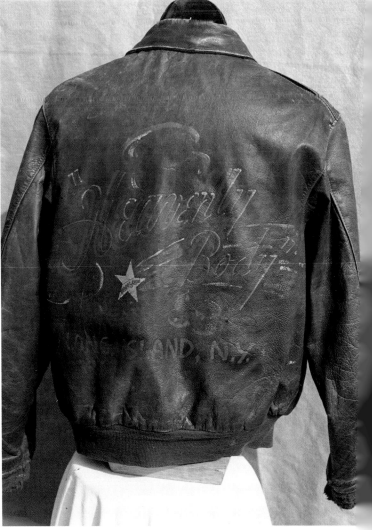

Front view of "Heavenly Body" A-2 jacket of the 392nd Bomb Group. (JS Industries)

Rear view of "Heavenly Body" A-2 jacket, 392nd Bomb Group. (JS Industries)

Europe, Africa & the Mediterranean

G.H. ARMSTRONG

"Doc's Boy," G.H. Armstrong, flew 30 missions on a B-24 named "Puss-n-Boots." The plus sign on the tail of the B-24 in the painting would indicate Armstrong was in the 577th Bomb Squadron, 392nd Bomb Group, 8th Air Force. No other information is available on G.H. Armstrong, but his A-2 jacket is quite stunning!

Left: A-2 jacket named to G.H. Armstrong, 577th Bomb Squadron, 392nd Bomb Group, 8th Air Force. (Richard Peacher)

Left: Rear view of Armstrong's A-2. Falling bombs indicate 30 missions. (Richard Peacher)

Detail of a popular 8th Air Force motif featuring an airplane superimposed on the 8th Air Force insignia, in this case a B-24 named "Puss-n-Boots" of 577th Bomb Squadron, 392nd Bomb Group. (Richard Peacher)

Left: Detail of painted leather 392nd Bomb Group patch, leather name strip, and "Doc's Boy" painted under the name strip. (Richard Peacher)

711TH BOMB SQUADRON, 447TH BOMB GROUP

LT. THOMAS B. BURRELL

The 711th Bomb Squadron, 447th Bomb Group flew B-17's out of Rattlesden, England with the 8th Air Force. Lt. Burrell (right) was the bombardier on a Flying Fortress "The Squirming Squaw." His A-2 jacket has a nice example of a chenille 711th Bomb Squadron patch sewn to the left breast.

Right: A-2 jacket worn by Lt. Thomas B. Burrell, bombardier, 711th Bomb Squadron, 447th Bomb Group, 8th Air Force. (Michael J. Perry)

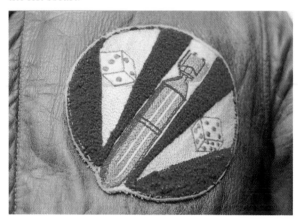

Left: Close up of chenille 711th Bomb Squadron patch on Burrell's A-2. (Michael J. Perry)

UNNAMED 711TH BOMB SQUADRON

There is no name other than "Belle" associated with this colorful A-2 jacket of the 711th Bomb Squadron. "Belle" could be a girlfriend or an airplane. Either way, the individual who thought up this design was very creative! The painting on the back shows a B-17 flying over a totally desolate Nazi Germany. Hitler's head is on a pole, a black cat scavenges through the trash, and a dog lifts his leg on the discarded Nazi flag. The right front panel has a painting of a bursting swastika with a "35" over it signifying 35 missions over Nazi territory. The left front has a painted 711th Squadron patch.

Back of 711th Bomb Squadron A-2 with a creative painting of a B-17 flying over a destroyed Germany and Hitler's head on a pole.

Left: Front of 711th Bomb Squadron A-2 with painted 35 mission insignia and 711th Bomb Squadron insignia.

THE 748TH BOMB SQUADRON, 457TH BOMB GROUP

RICHARD E. FITZHUGH

Richard E. Fitzhugh was the pilot of a B-17G "El Lobo II." He completed 30 missions with the 748th Bomb Squadron, 457th Bomb Group, 8th Air Force. In 1946, Fitzhugh piloted a B-17 which flew Churchill on a speaking tour of the U.S. and Cuba. "El Lobo II" went on to complete 100 missions with other crews and became the subject of a model kit. The jacket is one of the best encountered. The paint is as clean as if it were painted yesterday (it wasn't). The Captain's bars are painted directly on the epaulets, complete with highlights and shadows. The 457th became very well known as a group (The Fireball Group) and Fitzhugh removed his squadron patch and replaced it with the group insignia (hence the stitch holes showing under the group patch).

A-2 jacket of Richard E. Fitzhugh, 748th Bomb Squadron, 457th Bomb Group, 8th Air Force.

Crew of "El Lobo II," Richard Fitzhugh is second from left, back row. (Fitzhugh)

Back of Fitzhugh's A-2.

Right: Left shoulder of Fitzhugh's A-2. Note the painted Captain's bars with highlights and shading.

"KRAUT KRUSHER"

The individual who wore the "Kraut Krusher" is now unknown. We do know from the record left by his spectacular jacket that he flew 35 missions with the 457th Bomb Group. The red prop bosses on the B-17G in the painting identify the 748th Bomb Squadron and the markings place the airplane after the spring of 1944.

Right: Front of "Kraut Krusher" A-2 jacket, 748th Bomb Squadron, 457th Bomb Group, 8th Air Force. (JS Industries)

Below: 457th Bomb Group (The Fireball Group) insignia on left front panel of "Kraut Krusher" A-2. Stitch holes remain from where a leather name tag was sewn. (JS Industries)

Triangle "U" insignia of the 457th Bomb Group painted on right shoulder of "Kraut Krusher." (JS Industries)

8th Air Force insignia painted on left shoulder of "Kraut Krusher" A-2. (JS Industries)

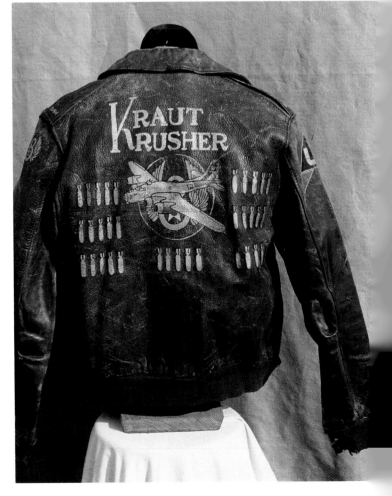

Right: Rear of "Kraut Krusher" A-2 jacket, 748th Bomb Squadron, 457th Bomb Group, 8th Air Force. (JS Industries)

Europe, Africa & the Mediterranean

826TH BOMB SQUADRON, 484TH BOMB GROUP

LT. J.H. SULLIVAN

Lt. J.H. Sullivan was a navigator with the 826th Bomb Squadron, 484th Bomb Group, 15th Air Force. The 484th flew B-24 Liberators out of Torretto Airfield, Italy and Casablanca, French Morocco. Sullivan's A-2 is adorned with classic tooled leather Italian made patches and insignia.

Tooled leather 15th Air Force patch affixed to left shoulder of J.H. Sullivan's A-2 jacket. (Arthur Hayes)

Left front panel of Sullivan's A-2 jacket with tooled leather name strip and navigator wing, and tooled leather 826th Bomb Squadron patch. Note, the bullets on the machine gun belt are grouped eight, two, and six. (Arthur Hayes)

Front view of Lt. J.H. Sullivan's A-2 flight jacket. (Arthur Hayes)

LT. ROBERT MORGAN

A second jacket (below) from the 826th was worn by Robert Morgan. Morgan was a B-24 pilot. The insignia on Morgan's jacket were made on the Isle of Capri and are of exceptionally fine quality.

Detail of tooled leather American flag right shoulder of Morgan's A-2.

Detail of tooled leather 826th Bomb Squadron patch made on the Isle of Capri. The 826th was designated a pathfinder unit, but did not serve in that function. "Morgan" is done in silver paint.

94TH FIGHTER SQUADRON

Detail of a 94th Fighter Squadron patch, as seen on the jacket in the photo.

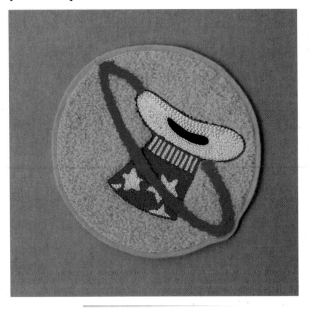

Left: This P-38 pilot wears an A-2 jacket with the patch of the famed "Hat in the Ring," 94th Fighter Squadron. His flight helmet is a British "B" pattern and the goggles are R.A.F. MK VII's. The 94th was assigned to the 1st Fighter Group, 8th and 12th Air Forces, and flew P-38 Lightnings from bases in England, Algeria, Tunisia, Sicily, Sardinia, and Italy. The P-38 behind him "Pitter Pat" was piloted by Lt. K.J. Sorace, 343rd Fighter Squadron, 55th Fighter Group. (Ilfrey via Campbell)

THE 364TH FIGHTER GROUP, 8TH AIR FORCE

The 364th Fighter Group, 8th Air Force flew P-38's and P-51's from Honington, England. The squadrons assigned to the 364th were the 383rd, 384th, and 385th.

GUY AUSTIN

Guy Austin served with the 385th Fighter Squadron, 364th Fighter Group, 8th Air Force. At the end of hostilities, Austin transferred to the 66th Troop Carrier Squadron in the States.

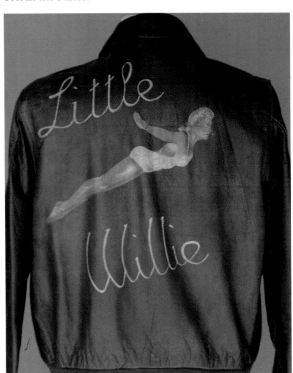

Above: A-2 of Guy Austin, 385th Fighter Squadron, 364th Fighter Group, 8th Air Force (later transferred to 66th Troop Carrier Squadron). Left: Back of Austin's A-2.

Trimmed bullion 8th Air Force patch on left shoulder.

Europe, Africa & the Mediterranean

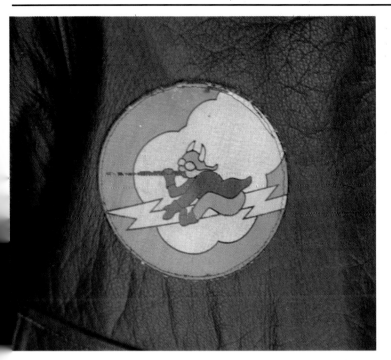

385th Fighter Squadron patch on right front of Austin's A-2.

66th Troop Carrier Squadron patch on left chest of Austin's jacket. Stitch holes remain from 364th Fighter Group patch, which was previously on the jacket. Name "Guy" is painted.

RICHARD C. COCHRAN

Flight Officer "Dick" Cochran was assigned to the 364th Fighter Group, 384th Squadron, of the 8th Air Force in April of 1945. Arriving late, Dick flew about 5 combat missions in the P-51 and remained with the occupation forces in Munich after the war. It's a mystery to Dick how he ever obtained this M-1941 field jacket but he well recalls painting the P-51 with "Kansas City Steak" nose art on the back.

Close up of the odd, color printed oilcloth 384th Fighter Squadron patch sewn to the left chest of Dick Cochran's M-1941 field jacket.

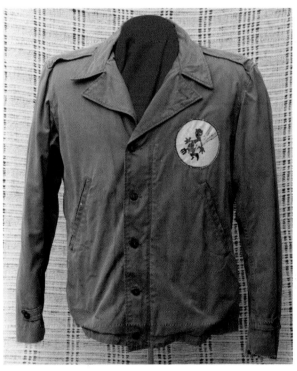

Left: Front view of Dick Cochran's Army issue M-1941 field jacket.

Below: Rear view of Dick Cochran's M-1941 field jacket.

Left: Close up of the silver finish P-51 Mustang "Kansas City Steak" hand painted by Flight Officer Dick Cochran on the back of his M-1941 field jacket. Note the worn unit tail code on rear stabilizer and the small red heart design below wording on nose.

370TH FIGHTER SQUADRON, 359TH FIGHTER GROUP

The 370th Fighter Squadron was assigned to the 359th Fighter Group, 8th Air Force. They were stationed at East Wretham, England and flew P-47's and P-51's. The original owner of this beautifully preserved A-2 is unknown.

Below: Detail of leather 370th Fighter Squadron insignia sewn to left breast of A-2 flight jacket. (Mirick)

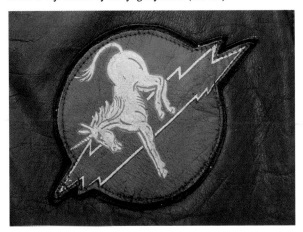

Left: Overall front view of 370th Fighter Squadron A-2 jacket. (Mirick)

512TH FIGHTER SQUADRON, 406TH FIGHTER GROUP

The 512th Fighter Squadron, 406th Fighter Group, was assigned to the 9th Air Force and served in England, France, and Germany. During the World War II years, the 512th flew many types of aircraft including A-20, A-24, A-25, A-26, A-35, A-36, UC-78, BC-1, P-39, P-40, and P-47. The war time mascot of the 512th, and central figure in the squadron insignia, was "Andy" the bulldog. Dr. Edgar A. Knowlton offered a nice explanation of Andy's history:

"I thought a brief summary regarding "Andy" the bulldog mascot of the 512th Ftr. Sqd. might be of interest. Perhaps many that knew Andy as our mascot might wonder what his eventual history was.

After Lt. Tomlinson, Andy's owner, was killed in action, Capt. Tom Wright, our personnel officer, and myself cared for him as much as we could. Andy had the run of the group area. He was fed all varieties of food (too many donuts). He developed severe eczema. Capt. Wright and myself took him to Brussels, Belgium to the veterinarian. After two weeks with the vet, we brought him home and his skin was clear. We were told his diet should be vegetables only.

At that time we were living in Nordholtz in a German family home. The lady was an excellent cook. They had a large vegetable garden. When it came time for us to leave for the USA, and because of Andy's allergy, we left Andy with them with their consent. I lost contact with the family so there was no follow up on Andy. This past year I made research as to Andy's future.

Jack Yarger informed me that he had some information and pictures of Andy from John E. Hagan. John was assigned to where the 406th had been. As John said, it was a mess after all the 406th left. Anyway, he states Andy had the run of the camp. Everybody fed him and petted him. Andy slept around John's quarters and on his Jeep. As to Andy's future after John returned home was, he probably died at the Camp.

I forgot to mention how Andy arrived in Europe. Lt. Tomlinson bought him in England. When it came time to move to the European continent, Lt. Tomlinson flew Andy on his shoulders to our station in France. It was quite a picture."

A young "Andy" with original owner Lt. Tomlinson (killed in action). (Michael J. Perry)

"Andy," in his prime, mascot of the 512th and Central figure in the squadron insignia. (Michael J. Perry)

Lt. Tomlinson. (K.I.A.), 512th Fighter Squadron, 406th Fighter Group, 9th Air Force, in fr[ont] of his P-47 "Razorback" with "Andy" on the cowling. (Michael J. Perry)

Europe, Africa & the Mediterranean

Detail of 512th Fighter Squadron insignia painted on Knowlton's jacket. The insignia features "Andy" wearing helmet and goggles, superimposed on a P-47 dropping three bombs. (Michael J. Perry)

Dr. Knowlton wearing his A-2. (Michael J. Perry)

DR. E.A. KNOWLTON

Captain "Doc" Knowlton was a flight surgeon who served with the 512th. His A-2 jacket has a painted version of the squadron insignia on the left front.

Below: A-2 jacket worn by Dr. E.A. Knowlton, 512th Fighter Squadron, 406th Fighter Group, 9th Air Force. (Michael J. Perry)

Detail of full color, leather 512th Fighter Squadron patch on Locke's jacket. (Michael J. Perry)

A-2 jacket of Major General John Langford Locke, 512th Fighter Squadron, 406th Fighter Group, 9th Air Force. (Michael J. Perry)

MAJOR GENERAL JOHN L. LOCKE

Major General Locke served in World War II, Korea, and was active during the Vietnam War. He was highly decorated and flew 126 combat missions. He was highly successful as a fighter pilot and as an administrator. General Locke was a graduate of West Point, class of 1941, and earned an MBA from Stanford. During World War II Locke flew a P-47 Thunderbolt with the 512th Fighter Squadron. Locke's A-2 jacket has a beautiful, full color version of the "Andy the Bulldog," 512th patch sewn on the left chest.

Right: Locke, standing in front of his Thunderbolt. (Michael J. Perry)

414TH NIGHT FIGHTER SQUADRON, 12TH AIR FORCE

The 414th Night Fighter Squadron, 12th Air Force was stationed at a number of airfields in Algeria, Tunisia, Sardina, Corsica, Italy, Belgium, and Germany. They flew a number of aircraft types, including A-20, P-70, Beaufighter, P-38, P-51, and P-61. The two men who wore the following pieces were P-61 Blackwidow pilots.

FRANK NORTHCUTT

Frank Northcutt's A-2 jacket was not available for photographing, but we do have a nice shot of him wearing it during the war and a beautiful example of a hand made 414th Night Fighter Squadron jacket patch.

Frank Northcutt, P-61 pilot with the 414th Night Fighter Squadron, 12th Air Force, wearing his A-2. As with many jackets from the Mediterranean, Northcutt's had an American flag on the right shoulder (visible in the photo). (Northcutt via Perry)

Above: Frank Northcutt in front of his P-61 "Hel'n Back." 414th Night Fighter Squadron. Left: Inside one of the 414th buildings, probably the officer's club or mess. Note the leaping black panther squadron insignias painted on the wall. (both - Northcutt via Perry)

Left: Frank Northcutt's 414th Night Fighter Squadron patch – a fantastic hand made example. (Michael J. Perry)

LOUIS J. PIMSNER

L.J. Pimsner was also a P-61 pilot with the 414th. His jacket has a painted example of a 414th insignia and a faded P-61 on the back.

Left below: A-2 jacket worn by L.J. Pimsner, 414th Night Fighter Squadron, 12th Air Force. Below: Back of Pimsner's A-2.

Painted 414th Night Fighter Squadron insignia and leather name tag "Louis J. Pimsner"

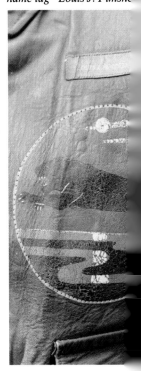

Europe, Africa & the Mediterranean

422ND NIGHT FIGHTER SQUADRON

The 422nd primarily flew P-61 Black Widows out of bases in England, France, Belgium, and Germany. They also used P-70, L-6, Oxford, C-78, and A-20 aircraft types.

Two 422nd Squadron members in front of a P-61 Black Widow "Shoo-Shoo Baby." The Lieutenant wearing the A-2 jacket and overseas hat has a 422nd Squadron patch on the left chest of his A-2. The cloth jacket worn by the other man is a B-10. (Campbell)

LtCol Oris B. Johnson of the 422nd Night Fighter Squadron wearing a B-10 jacket with the squadron patch clearly visible on the left chest. (Campbell)

9TH TROOP CARRIER COMMAND B-15 JACKET

The practical cotton B-15 flight jacket was preferred by many airmen due to its light weight, warmth and comfort. This jacket, with its simple hand colored canvas insignia, is an interesting example worn by someone attached to the troop carrier command that liberated Europe via every major airborne operation from D-day on.

Front view of the 9th Troop Carrier Command B-15 jacket. Note that it is a relatively early model with no modifications.

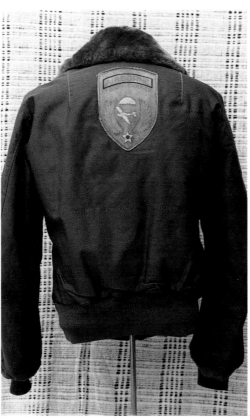

Rear view of the B-15 jacket with large 9th Troop Carrier Command insignia applied to the back.

Close up of the factory printed Army Air Forces insignia on the left shoulder of the 9th Troop Carrier Command B-15 jacket. This "factory" insignia was often short lived due to day to day wear and gasoline "dry cleaning" methods employed in the field.

Close up of the unusual hand painted on canvas 9th Troop Carrier Command insignia sewn to the upper back of this AAF issue B-15 flight jacket. Note also the rich, reddish-brown fur collar that was typical of the B-15 jacket.

GENE ARNOLD

Lt. Gene Arnold was a pilot with the 860th Bomb Squadron, 493rd Bomb Group, 8th Air Force. Arnold flew 33 missions. The 493rd flew out of Elveden Hall and Debach, England. Arnold's A-2 jacket has a nicely painted squadron patch on the left breast and a nude leaning on a bomb in front of the "winged 8" painted on the reverse.

Below left: Front view of Gene Arnold's A-2 flight jacket.
Below: Rear view of Arnold's A-2. (Mick Prodger)

Detail of 860th Bomb Squadron insignia painted on the breast of Arnold's A-2. (Mick Prodger)

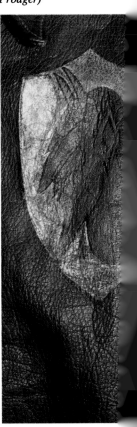

P-51D "Nancy Lee" flown by R.L. Baggett. (Baggett via Jones)

RICHARD L. BAGGETT

Richard L. Baggett flew P-38's and a P-51D Mustang named "Nancy Lee" with the 402nd Fighter Squadron, 370th Fighter Group, Ninth Air Force. The 402nd was stationed at several airfields in England, France, Belgium, and Germany. At the end of the war they were at Fritzlar, Germany. Baggett's jacket has an unusual example of a personal patch with a P-38 wearing boxing gloves, painted on a leather disc with his name embossed. On the back of the jacket is the 402nd Squadron patch which was painted long after WWII. Of special interest is the plastic "Dingy Wistle" at the collar of the jacket, which is clearly visible in the war time photographs.

Left: Lt. R.L. Baggett – Clearly visible are the P-38 Patch and dingy whistle seen in the color photos of this jacket. (Baggett via Jones)

Front view of A-2 jacket worn by R.L. Baggett, 402nd Fighter Squadron, 370th Fighter Group, Ninth Air Force. (J. Jones)

Rear of Baggett's A-2 jacket. Squadron insignia was painted post-war by Baggett's grandson. (J. Jones)

Left: Detail of personal patch of a P-38 Lightning on its tail with boxing gloves, and "Richard L. Baggett" embossed in the leather. (J. Jones)

CLAUDE A. "CHUCK" BERRY – 91st TROOP CARRIER SQUADRON

Chuck Berry has the unique honor of being a "3 war" Army aviator. As a glider pilot in WWII he flew with the 91st Troop Carrier Squadron of the 439th T.C. group and was one of 72 glider pilots to participate in Operation "Repulse" – the aerial relief of Bastogne. Chuck flew missions into Holland and Germany and went on to fly liaison planes in Korea and helicopters in Vietnam! Chuck's old A-2 saw a lot of service time and was worn for years after the war in his civil aircraft and also on trips and reunions! It has been relined, all knit material and zipper have been replaced but the leather is still good and solid and the artwork remains a testimony of participation in the ETO.

Right: Front of Chuck Berry's "91st Troop Carrier Squadron" A-2 jacket. The well worn decal on leather 1st Troop Carrier Command patch can be seen on the right chest. Far right: Rear view of "91st Troop Carrier Squadron" A-2 jacket. Names of operations the 91st participated in with gliders can be seen in the centers of the flak bursts – Chuck participated in the three listed across the top.

"BLOND BOMBER"

The B-10 Army Air Force flight jacket was the first cloth jacket to see widespread issue and service during the war. This particular example appears to have been issued to a 15th Air Force air gunner who was fond of blonds! The jacket is well preserved and shows none of the typical fraying and damage common to cloth, combat tested jackets. The artwork is outstanding and has added a very personal element to an otherwise business-like garment.

Rear view of the "Blond Bomber" B-10 jacket. It's interesting to note that the original title was "Blond Bombers" but someone has painted out the "s." This is not as glaringly evident when the jacket is viewed, in normal lighting conditions, with the naked eye.

Left: Close up of the beautifully stylized, hand painted air gunner wing on the left chest of the "Blond Bomber" B-10 jacket.

Front view of the "Blond Bomber" B-10 flight jacket.

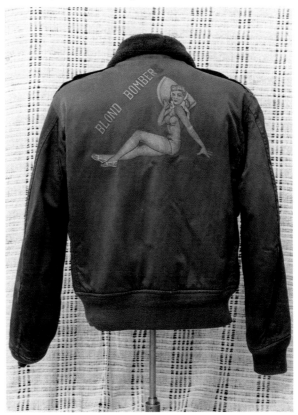

Close up of the 15th Air Force patch on the left shoulder of the "Blond Bomber" B-10. A faint trace of the words "Army Air Force" below indicates that this insignia was painted using the factory printed A.A.F. patch design as a guide.

Europe, Africa & the Mediterranean

CHARLES W. BLOUNT

Although not actually a flight jacket, the C-2 vest was used as a flying outer garment during the second world war. This versatile garment was intended to be used as an element of the layering principal for warmth in varying degrees. The example shown here was issued to Flight Officer Charles W. Blount while awaiting overseas movement after his assignment to the 314th Troop Carrier Group at the Sedalia army air base in 1943. While stationed in North Africa, Chuck participated in ferrying operations in WACO CG-4A gliders in preparation for the invasion of Sicily. Chuck stated that this light, knit wool sweater was often perfect for the wide variety of temperatures experienced in the area and he wore it often while flying. The leather name tag is strong evidence that it was indeed worn as an outer garment. Chuck retained his assignment to the 314th Troop Carrier Group, 61st Squadron and participated as a glider pilot on the airborne assaults into Holland and Germany.

Period photo of a B-25 crew somewhere in North Africa. The man third from the left is wearing a C-2 vest. Technically a sweater, the C-2 vest was a versatile garment. The insignia on the fuselage is that of the 94th Bomb Squadron.

Detail of the impressed leather name tag sewn on the left chest of Flight Officer Blount's C-2 vest. This was recommended procedure for in flight garments during the war.

Group of personal items from the war-time career of Flight Officer Charles Blount. Of interest are the various sleeve insignia, wings, rank insignia and collar devices. Note the used leather name tags as worn on flight garments.

WWII Army Air Force issue C-2 vest issued to Flight Officer Chuck Blount. Flight Officer Blount was a glider pilot with the 314th Troop Carrier Group and often wore this sweater as an outer garment while flying in intermediate temperatures.

Detail of the nomenclature label sewn inside the neck of Flight Officer Blount's C-2 vest.

RICHARD L. BOWLING

Dick Bowling joined the Army in April of 1941 and was assigned to the headquarters of the 51st Pursuit Group in communications capacity. He was accepted as a candidate for Staff Sergeant pilot training in late 1941 but washed out because his check instructor felt he was "dangerous and erratic." He later applied for glider pilot training and went overseas with the 61st squadron of the 314th Troop Carrier Group in May of 1943. Dick participated in the airborne operation into Holland on September 18, 1944 and returned home on July 3, 1945. Dick Bowling's A-2 jacket was hand painted by a squadron mate while overseas.

Below left: Front view of Dick Bowling's A-2 jacket. Note aircraft national insignia marking on CG-4A glider fuselage in background. Below right: Rear view of Dick Bowling's A-2 jacket. Note hand painted likeness of CG-4A glider with Bowling's wife's name "Bette" painted on the nose section. Background is a rendition of the aircraft national marking mixed with a few "hostile" flak bursts! (Silent Wings Museum)

EUGENE BRAUCHER

Lt. Eugene Braucher served as a navigator on a B-17G, "Big Dick," with the 422nd Bomb Squadron, 305th Bomb Group, 8th Air Force. The scoreboard on Braucher's jacket indicates 22 missions. Braucher's A-2 is one of the most decorated jackets encountered.

Front view of Lt. Eugene Braucher's A-2 jacket. (Michael J. Perry)

Rear view of Lt. Eugene Braucher's A-2 jacket. (Michael J. Perry)

Left: Right breast of Braucher's A-2 with painted 22 mission scoreboard and 305th Bomb Group patch affixed. (Michael J. Perry)

Europe, Africa & the Mediterranean

B-17F with very similar nose art to the B-17G on which Lt. Braucher served. (John Campbell)

Left breast of Braucher A-2 with leather embossed navigator wing, name strip, and squadron patch. (Michael J. Perry)

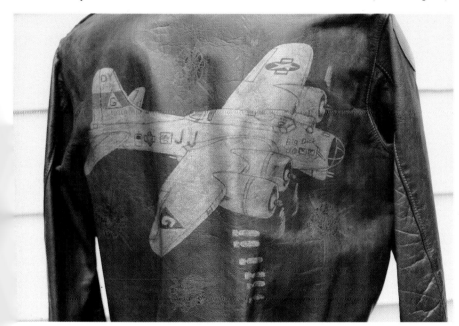

Close-up of B-17G dropping bombs in heavy flak painted on the back of Eugene Braucher's A-2 jacket. (Michael J. Perry)

Left: Lt. Eugene P. Braucher, 422nd Bomb Squadron, 305th Bomb Group, 8th Air Force. The reverse of this photo is inscribed, "Front view of jacket. Note – 22 bombs, one for each mission. 1945-England –right after I got my A-2 jacket." Right: Back of Lt. Braucher's A-2 jacket. Photo reads "Rear view of my leather flying jacket. Some stuff, huh! Six bombs, one for each lead." (Braucher via Perry)

Left shoulder of Braucher's A-2 jacket with leather 8th Air Force patch and rank insignia. (Michael J. Perry)

Right shoulder of Braucher's A-2 jacket with leather 9th Air Force patch and rank insignia visible. (Michael J. Perry)

ROBERT E. BUIS

Staff/Sgt. Robert Buis flew 35 missions as tail gunner on a B-17 "Sunday Punch" with the 384th Bomb Group, 8th Air Force. The red cowlings on the B-17 in the painting likely indicate the 546th Bomb Squadron, as the 384th Group began using colored engine cowlings to designate squadrons from April 1945. The 384th flew out of Grafton Underwood, England.

Right: A-2 jacket "Sunday Punch" worn by Staff/Sgt. Robert E. Buis, 384th Bomb Group, 8th Air Force. The art work represents 35 missions on a B-17 G.

T. JAMES W. CAMPBELL – 1st TROOP CARRIER COMMAND

Lt. Jim Campbell graduated from 29 Palms Glider School in January of 1943 and went to the European Theatre of Operations (E.T.O.) with the 79th Squadron of the 436th Troop Carrier Group. Jim was "D" flight leader for the 79th Squadron on D-day, June 6, 1944 and left England for Normandy at 3:30 a.m. that morning. Jim volunteered for a five glider resupply mission on D-day + 6 (June 12- 5p.m.) and also participated in the Market-Garden operation into Holland, and the Varsity Mission across the Rhine. He received the Air Medal with two clusters, the Bronze Arrowhead and the Distinguished Unit Badge for his service as a combat glider pilot in WWII. He died May 8, 1988. His old A-2 leather flying jacket was extremely well kept and remained a valuable memento of service days until his death.

Below: Front view of Lt. Jim Campbell's A-2 jacket.

Lt. James W. Campbell on an English airfield in the summer of 1944. His trusty A-2 jacket can be seen draped casually over his left arm. The dark fuselage of a British Horsa glider fills the background.

Left breast of Lt. Jim Campbell's A-2 jacket. The leather strip is impressed "J.W. Campbell." Note the decal on leather 1st Troop Carrier Command patch immediately below name

Europe, Africa & the Mediterranean

"THE CAROLINIAN"

Tail markings on the B-26 Marauder and the same yellow and black stripes around the squadron insignia identify "The Carolinian" with the 387th Bomb Group, 9th Air Force. The 387th operated from air fields in England, France, and Holland. The art work on this A-2 is of exceptional quality and is in remarkable condition.

Left: A-2 jacket of the 387th Bomb Group, 9th Air Force "The Carolinian." Clearly visible is the yellow and black diagonal stripe tail marking of the 387th Bomb Group. The nose art on the airplane shows the Carolina moon shining on pine trees. Far left: Beautifully painted squadron insignia with group markings of yellow and black stripes surrounding, on "The Carolinian." (Manion's)

JOSEPH CARUANA

Captain Joseph Caruana flew P-51 Mustangs with the 2nd Fighter Squadron, "American Beagle Squadron," 52nd Fighter Group, 15th Air Force. Caruana's jacket is plain with the exception of a nicely tooled leather squadron patch, likely made on the Isle of Capri. Also, at the collar is a small bell often purchased by American airmen as a souvenir from the Isle of Capri.

"American Beagle Squadron" insignia painted on the nose of a Spitfire. The squadron flew Spitfires from 1942 to 1944, when they switched to P-51s. Obviously a spoof on the famed American Eagle Squadron, which was made up of American volunteers in the Royal Air Force in the early stages of WWII. The Eagle Squadrons later became the 4th Fighter Group in the USAAF. (John Campbell)

"American Beagle Squadron" insignia painted on a wall, probably in an officer's club. (Caruana via Perry)

Front view of A-2 jacket worn by Captain Joseph Caruana, 2nd Fighter Squadron, 52nd Fighter Group, 15th Air Force. Note the small silver bell at the collar from the Isle of Capri. (Michael J. Perry)

American Flight Jackets, Airmen & Aircraft

2nd Fighter Squadron insignia in tooled leather sewn to left breast of Caruana's A-2. (Michael J. Perry)

Captain Joseph Caruana in the cockpit of his P-51 Mustang. (Caruana via Perry)

Captain Joseph Caruana wearing his A-2 jacket with a lovely Companion. (Caruana via Perry)

J.W. CHRISTIAN

Lt. Christian was a navigator with the 97th Bomb Squadron, 47th Bomb Group, 12th Air Force. The name on his jacket indicates he went by "Chris." The 97th flew A-20 Havocs and A-26 Invaders out of French Morocco, Tunisia, Algeria, Malta, Sicily, Italy, Corsica, and France. Christian's A-2 has all insignia painted on, including 97th Bomb Squadron patch, name, navigator's wing, and American flag (commonly seen on 12th Air Force jackets).

A-2 jacket of Lt. J.W. Christian, navigator, 97th Bomb Squadron, 47th Bomb Group, 12th Air Force. (JS Industries)

Detail of painted insignia on Christian's A-2. Squadron patch is the 97th Bomb Squadron. (JS Industries)

Europe, Africa & the Mediterranean

FRED J. CHRISTENSEN, JR.

Nicknamed "Rat-Top," Fred Christensen was the first 8th Air Force Fighter Pilot to shoot down six enemy aircraft on a single mission. Christensen was a natural at air combat; he was assigned to the famed 56th Fighter Group ("The Wolf Pack") in August of 1943 and he was an ace by February of the following year. An ace is any pilot who claimed five or more air to air victories over enemy aircraft. A four-time ace, Christensen finished the war as one of the 8th Air Forces' top aces with 21-1/2 aerial victories. His personalized A-2 leather jacket and flight helmet with original goggles remain silent witnesses to those turbulent hours in the air that placed Christensen in a realm few fighter pilots ever reach.

Photo shows A-2 jacket, AN-H-15 summer flying helmet, B-8 goggles and portrait photo from the wartime flying career of four-time ace Fred J. Christensen, Jr.

Front view of Fred Christensen's A-2 jacket.

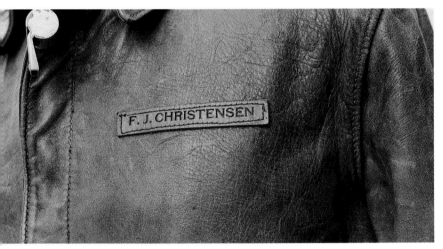

Close up of the left chest of Christensen's A-2 jacket bearing the impressed leather name tag. Of special note is the nickel plated "police" whistle attached to the throat closure hook at the collar. These whistles were packed in aircraft survival kits to provide a source of noise for would-be rescuers to seek out in the event the airman was forced down.

Close up of the neck area of Christensen's A-2 jacket revealing hand inked name and Army Air Force specification label in lining.

Close up of Army Air Forces decal on the left shoulder of Christensen's A-2 jacket.

J.H. COBB

J.H. Cobb's A-2 jacket provides a nice example of a flight jacket used by non-flying personnel. Cobb was a member of the 82nd Airborne Division and he traded a pair of jump boots for the jacket.

Front view of A-2 jacket worn by J.H. Cobb, 82nd Airborne Division. (Mick Prodger)

Lt. J.H. Cobb, 82nd Airborne Division wearing his non-regulation A-2 jacket. (Cobb via Prodger)

Detail of J.H. Cobb name in leather and 82nd Airborne Division patch. (Mick Prodger)

LEE E. DANO

Lee E. Dano served with the 15th Air Force. Unfortunately, that is all the information available. We can speculate from the well executed B-17G painting on the back of his A-2 jacket that he was a member of a B-17G crew. The painting is of good quality and typical of the artwork done in Italy and on the Isle of Capri. Although worn, the painting remains vivid – a simple but classic A-2.

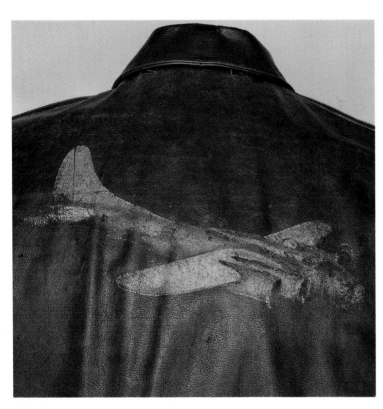

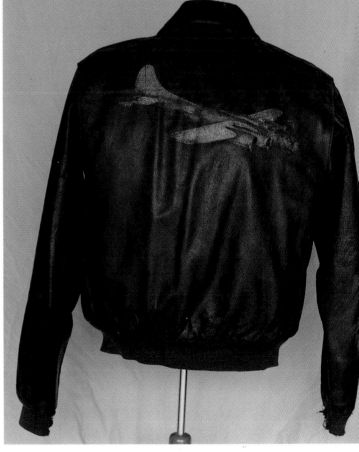

Above: Rear view of A-2 jacket worn by Lee E. Dano, 15th Air Force. Aircraft is a B-17G. Left: Detail of B-17G painting on Lee Dano's A-2 jacket. (Mick Prodger)

Europe, Africa & the Mediterranean

SGT. ODES DRENNAN

Odes Drennan served on a B-24 Liberator with the 744th Bomb Squadron, 456th Bomb Group, 15th Air Force, at Cerignola, Italy. Sgt. Drennan had three unusual flight jackets. The first is an A-2 jacket with a white fur collar attached. The front of the jacket has tooled leather group and squadron patches, as well as leather flashes reading "Pro Deo Et Patria Sempre" and "Odes D. Drennan Big "D" Tex." The jacket also has shoulder insignia and a nicely painted B-24 "Salvo Susie" on the reverse.

One crew assigned to "Salvo Susie" in front of the aircraft. (Drennan via Perry)

Detail of "Salvo Susie" B-24 painting on back of Drennan's A-2. (Michael J. Perry)

Above: Front view of Sgt. Odes Drennan's A-2 jacket with fur collar added. Below: Rear view of Sgt. Drennan's A-2. (Michael J. Perry)

Far left: Sgt. Odes Drennan wearing his A-2 jacket before it was "completed." Note it does not yet have the group patch or the fur collar in this photograph. Left: Sgt. Odes Drennan (center) and two other crewmen in front of "Salvo Susie." (Drennan via Perry)

Left: Right breast of Drennan's A-2 jacket with tooled leather 456th Bomb Group insignia and "Pro Deo Et Patria Sempre" tab in place. (Michael J. Perry)

Right: Tooled leather 744th Bomb Squadron patch and Odes D. Drennan Big "D" Tex name strip attached to left breast of Drennan's A-2 jacket. (Michael J. Perry)

Right: Left shoulder of Drennan's A-2 with hand painted American flag. (Michael J. Perry)

Left: Hand tooled leather 15th Air Force patch affixed to right shoulder of Drennan's A-2. (Michael J. Perry)

The second jacket from Sgt. Drennan is a privately purchased brown leather jacket with a white fur collar. The reverse has a nicely done painting of a lady and on the left shoulder is a 15th Air Force patch. The painting is of quality consistent with Italian jacket art.

Right: Rear view of privately purchased jacket of Sgt. Odes Drennan. (Michael J. Perry)

Left: Hand painted 15th Air Force patch on left shoulder of Drennan's jacket. (Michael J. Perry)

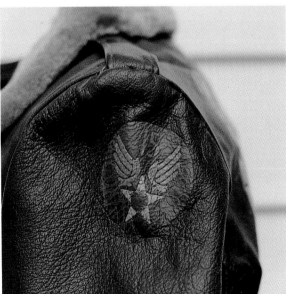

Europe, Africa & the Mediterranean

Nose art on B-24 Liberator "Miss Laid," 456th Bomb Group, 15th Air Force. This provides a good example of nose art being duplicated on a flight jacket. (Drennan via Perry)

The third jacket, which came from Sgt. Drennan, is named to G.K. Richardson. Richardson served on a B-24 "Miss Laid." The flying horse patch is an unofficial 456th Bomb Group insignia for "Steed's Flying Colts" – named for Colonel Thomas W. Steed, the commanding officer. The mission scoreboard indicates 30 missions.

Unofficial 456th Bomb Group insignia – "Steed's Flying Colts." (Michael J. Perry)

Above: Front view of Richardson's A-2 jacket, 456th Bomb Group, 15th Air Force. Below: Rear view of Richardson's A-2 jacket, 456th Bomb Group, 15th Air Force. The style and quality of the painting is very typical of the work done by Italian artists. (Michael J. Perry)

Left: Army Air Force insignia on left shoulder of Richardson's A-2. (Michael J. Perry)

CYRIL A. DWORAK – "WINN'S WARRIORS"

Staff Sgt. Cyril Dworak survived 233 hours of combat flying as an air gunner with the 96th Bomb Group of the 8th Air Force while stationed at Snetterton Heath, England from September of 1944 to April of 1945. The 96th Bomb Group became known as the "Snetterton Falcons" and enjoyed an illustrious reputation in the annals of air combat in the E.T.O. Dworak flew his entire combat tour of thirty missions with one crew and dubbed his A-2 jacket "Winn's Warriors" in honor of his pilot. He has the distinction, as an air gunner, of one confirmed enemy fighter "kill" on the April 7, 1945 mission against Kaltenken, Germany.

Above: Front view of Cyril Dworak's A-2 jacket.
Below: Rear view of Cyril Dworak's A-2 jacket.

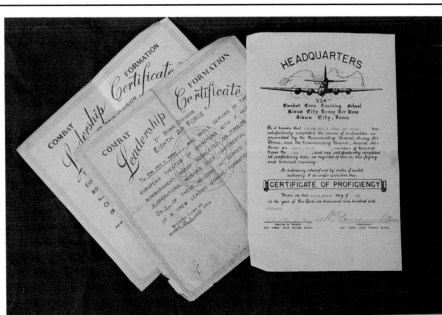

Certificates earned by Cyril Dworak for proficiency in combat crew training and participation with the lead crew in a 3rd bomb division combat formation.

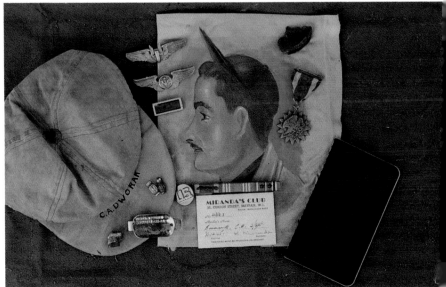

Group of personal items from the wartime career of Cyril Dworak. The "Black Book" in the lower right records all of Dworak's combat missions with a special note on the day he scored a confirmed victory over an attacking German fighter. Air crew and air gunner wings, an air medal, ribbon bars, dog tag, London club card, "ball cap," and hand painted portrait of Dworak are also part of this interesting group of artifacts that often accompany a flight jacket.

Detail of Dworak's initials "CAD" hand painted on the right chest of his A-2 jacket. Leaving the periods out was no mistake for CAD was a popular term for a "man of the world" at the time.

Europe, Africa & the Mediterranean

Detail of the hand embroidered felt 338th Bomb Squadron patch sewn to the left chest of Cyril Dworak's A-2 jacket. A close examination reveals traces of a painted 339th Squadron insignia below the 338th patch. Both squadrons were components of the 96th group and it was not uncommon for airmen to be shifted from squadron to squadron to compensate for combat losses or stateside rotation or personnel who had completed their combat tour.

Detail of hand painted 8th Air Force patch design on the left shoulder of Cyril Dworak's A-2 jacket. Note this is the final pattern of the 8th Air Force patch with long, upswept wings. The yellow border was one of many unofficial ways to denote the wearer was a member of a lead crew.

Detail of beautiful "Winn's Warriors" motif hand painted by Joe Bodner (also of the 96th Bomb Group) on Cyril Dworak's A-2 jacket. This artwork is exceptional for the period; an experienced artist, Bodner has employed much life-like detail and a fairly wide color scheme. Of special note is the silhouette of a fighter plane superimposed on the swastika; this denotes Dworak's victory as an air gunner over an attacking German fighter on the April 7, 1945 mission against Kaltenken, Germany. Also noteworthy are the thirty mission bombs in a rather casual configuration below the artwork. The paint pigment has reacted in an unusual way on the leather, creating an almost "embossed" appearance on all painted surfaces.

"FINAL APPROACH"

Five crew members of "Final Approach" B-24J pose in front of the aircraft nose art after completing 30 missions with the 752nd Bomb Squadron, 458th Bomb Group, 8th Air Force, out of Horsham St. Faith, England. This photo provides a great example of nose art being carried over into jacket art. The painting shows a B-24 on "Final Approach" to the U.S.A. This aircraft was hit by flak and crashed in Lechfeld, Germany, 8 April 1945 while on its 123rd mission. (Hetzel via Campbell)

"DUKE"

This American airman, "Duke," served with the 95th Bomb Group, 8th Air Force. The 95th flew B-17s out of Framlingham and Horham, England. Duke's A-2 has his name and the 95th Bomb Group insignia painted on the left chest and a pin-up girl with "What's Cookin'" on the back.

Below: Front and back of 95th Bomb Group, 8th Air Force, A-2 jacket named only to "Duke." (JS Industries)

JOSEPH P. FORD

1st Lt. Joseph P. Ford served as a navigator/bombardier with the 756th Bomb Squadron, 459th Bomb Group, 15th Air Force. The 459th served at Giulia Airfield in Italy and flew B-24s.

Right: A-2 flight jacket worn by 1st Lt. Joseph P. Ford, 756th Bomb Squadron, 459th Bomb Group, 15th Air Force. The patch on the left front panel is the 756th Bomb Squadron and the left shoulder has a 15th Air Force patch painted on. Left: Rear view of Ford's A-2 jacket with 50 bomb mission scoreboard painted on. (Martin Callahan)

Europe, Africa & the Mediterranean

"FREE LANCE"

"Free Lance" could be the nickname of an individual or of an airplane. This A-2 jacket has a nice clear painting of a B-17G dropping bombs on the swastika of Nazi Germany. This jacket is from the 8th or 15th Air Force.

Below: A-2 jacket "Free Lance" of the 8th or 15th Air Force. Left: Detail of painting depicting a B-17G dropping bombs on a swastika. (JS Industries)

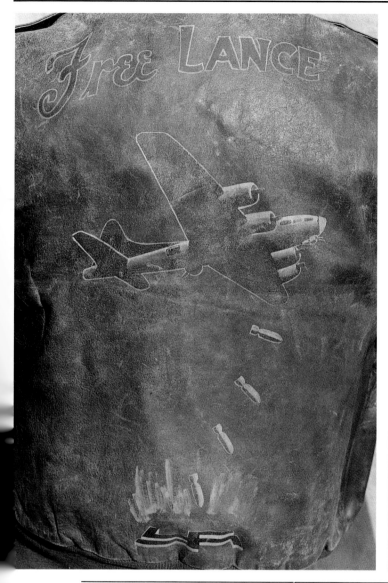

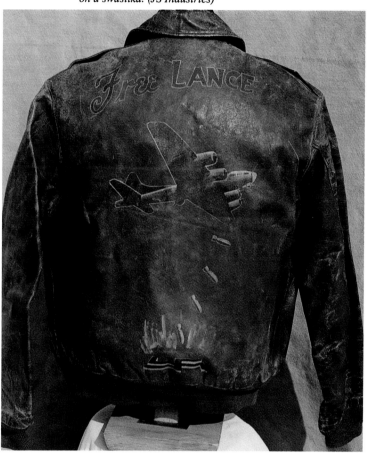

Left: Rear view of A-2 flight jacket worn by S/Sgt. James F. Grumann, 613th Bomb Squadron, 401st Bomb Group, 8th Air Force. The painting features a 30 bomb mission scoreboard and "Wimpy" from the Popeye cartoon. (Mick Prodger)

JAMES F. GRUMANN

S/Sgt. James F. Grumann was the ball turret gunner on a B-17, 613th Bomb Squadron, 401st Bomb Group, 8th Air Force. The 401st was stationed at Deenethorpe, England. The pilot of the plane was Captain McGoldnck, hence the name "Mac's Goldbricks." Grumann survived 30 missions, but was badly wounded on his 30th. The painting on the back, although worn, features "Wimpy" from the Popeye cartoon strip.

Right: Right front panel of Grumann's A-2 jacket with hand painted 613th Bomb Squadron insignia. (Mick Prodger)

ROBERT W. GLAESER

The 452nd Bomb Group was stationed at Deopham Green, England. "Doc's Baby" was a B-17 of the 452nd Bomb Group, 731st Squadron, 8th Air Force. Daniel E. "Doc" Hannon was the pilot. T/Sgt. Robert W. Glaeser was the engineer/gunner on the crew from the time it was formed in 1944 until June of 1945 when they all came home. Glaeser's jacket is a privately purchased leather jacket. Glaeser recalls that "six of the fellows on Hannon's crew bought them from a guy on our base. They cost five pounds (English money). The pound was worth $4.68 American money, so the jacket cost us about $25.00 – lots of money in 1945. The fellow that did the art work is a fellow named Dave Roberts. He lives in Lake Forrest, Illinois – saw him at the reunion in Des Moines..."

Front of "Doc's Baby." Privately purchased flight jacket of R.W. Glaeser. The leather squadron patch is the 731st Bomb Squadron. Tied on the right epaulet is a piece of parachute cord. (Glaeser)

T/Sgt. Robert W. Glaeser, engineer/gunner on "Doc's Baby," B-17, 452nd Bomb Group, 731st Squadron, January 1945. (Glaeser)

Rear of "Doc's Baby" with painted character and 35 mission scoreboard. (Glaeser)

Right: Detail of leather 731st Bomb Squadron insignia. (Glaeser)

Left: Left shoulder of "Doc's Baby" with American, British, Russian, and French flags affixed. (Glaeser)

Europe, Africa & the Mediterranean

RALPH JENKS

Ralph Jenks served as a pilot in the 310th Ferrying Squadron, 31st and 27th Air Transport Groups, 302nd Transport Wing, United States Strategic Air Force (USSTAF). The 302nd was at a number of air fields in England and France. Due to the nature of ferrying duty, Ralph has flying time in almost every type of WWII operational aircraft. Ralph Jenks' A-2 jacket is unique in that it has had a fur collar added. Unfortunately, the squadron patch (visible in the group photo) has been removed.

Ralph Jenks in the cockpit of a P-38 Lightning.

A-2 jacket worn by Ralph Jenks. Note added fur collar.

Left: Ralph Jenks, first from left, along with other 310th pilots, wearing his A-2 jacket. Fourth man from left also has added a fur collar to his A-2 jacket. (Ralph Jenks)

ROBERT S. JOHNSON – AN ACES' B-15 FLIGHT JACKET

Unlike many of his contemporaries, Bob Johnson seemed to prefer the plain, practical cloth B-15 jacket over the classic A-2 leather jacket. Johnson can be seen in many period photos wearing what appears to be this B-15 jacket, or one like it. Due to the fact that his combat flying career of 91 missions lasted from only January of 1943 to May of 1944 it is unlikely that he owned more than one of these jackets. Bob Johnson scored 28 air victories while flying with the famous 56th fighter groups' "Wolf Pack." He passed this B-15 jacket on, along with a pair of his English flying gloves, to an admirer after the war. The historical significance of these types of artifact often overcome their lack of eye appeal.

Front view of WWII Fighter Ace Robert S. Johnson's Army Air Force issue B-15 flight jacket. Preferred by Johnson over the A-2 jacket, this B-15 is plain and very simple in appearance. A personal letter from Johnson confirms that he flew some of his 91 combat missions in this jacket.

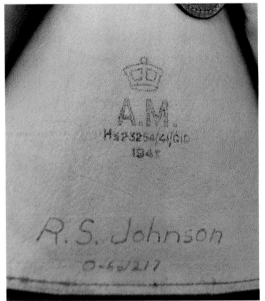

Left: Personal markings inside one of Robert S. Johnson's British issue leather flying gauntlets.

This photo shows a small group of artifacts passed on to an admirer by Johnson after the war. Ironically, the autographed photo shows Johnson in the cockpit wearing an A-2 jacket. His B-15 jacket and British R.A.F. issue flight gauntlets can also be seen.

DON KNIGHT
Don Knight flew B-17s in the 615th Bomb Squadron, 401st Bomb Group. The airplane, "Mary Alice", was named for his mother. The lettering at the bottom of the jacket, which is not fully visible, reads, "Gnatzi-Knightmare." (Colones)

Europe, Africa & the Mediterranean

J.E. LABEAU

J.E. Labeau was an air crewman on an 8th Air Force B-17 named "Montezuma." That is where the name "Monty" on the back of his A-2 originated. It is unknown what squadron or group he served in. There are 29 bombs falling on the swastika in the painting, probably indicating as many missions.

FREDERICK C. LEROY – "CRANKY YANKEE"

Sgt. Fred Leroy was a flight engineer/gunner on B-17Gs of the 487th Bomb Group, 839th Squadron of the 3rd Bomb Division, 8th Air Force. While stationed at Lavenham, England, Fred flew 35 missions from December of 1944 to April of 1945. The most memorable were the two trips to Berlin. Fred's jacket bears his crew given nickname "Cranky Yankee" which reflected his eastern roots of Pennsylvania. He recalls that every member of his crew had a motif and nickname painted on their jackets that represented their home state or region. This jacket shows very little of its years of age and left Fred's possession only when the long trip to Florida upon retirement loomed ahead.

W.E. LUCAS

Very little information is available on the Lucas A-2 jacket. It is unusual in that it is made of goat skin, as opposed to cow or horse hide. The jacket features an unidentified squadron or group insignia with a four engine bomber silhouetted over a bomb. The aircraft is a bit nondescript, but probably represents a B-17. The patch is tooled leather of the style commonly seen on 15th Air Force jackets.

A-2 jacket of J.E. Labeau with cloth aircrew wing sewn over leather name strip.

Rear view of Fred Leroy's "Cranky Yankee" A-2 jacket. Relatively simple artwork is typical of many A-2s but 35 bombs represents lots of time on the job, no matter how they're painted! Note the bombs at 10 and 1 o'clock bear small red "B"s for Berlin. The letters and alternating stripes of the shield were once white. The failing pigment in some way affected the dye in the leather, leaving these areas a tan shade.

A-2 jacket worn by W.E. Lucas, most likely a 15th Air Force jacket.

Left: Back of Labeau's A-2 jacket with painting depicting the 8th Air Force insignia over 29 bombs falling on the swastika insignia of Nazi Germany. The name "Monty" is short for Montezuma, the nickname of the B-17 on which Labeau served.

"LUFTWAFFE" A-2's and B-3's

Right: This B-17F, "Miss Quachita", 323rd Bomb Squadron, 91st Bomb Group, was shot down by Major Heinz Bär, Kommandeur of II/JG 1 in February of 1944. Here Major Bär and wingman Oberfeldwebel Leo Schumacher, inspect the downed aircraft. Bär is wearing an A-2 jacket, to which he has added an Iron Cross First Class and Luftwaffe shoulder boards. Schumacher is wearing a B-3. (via Campbell Archives)

Unteroffizier Heinz Hanke wearing an A-2 jacket from an American bomber crewman who completed 18 missions on "Miss Behavin." (Eric Mombeek)

Major Bär (far right), wearing his A-2 jacket, gives direction to his pilots. Lt. Wegner, fifth from left, is wearing an American B-3 jacket. (Eric Mombeek)

Heinz Bär (wearing A-2) and Schumacher (wearing B-3) exploring the wreckage of "Miss Quachita." (Eric Mombeek)

Europe, Africa & the Mediterranean

DON A. LUTTRELL

Lt. Don A. Luttrell was a P-38 pilot with the 49th Fighter Squadron, 14th Fighter Group, 15th Air Force. The 49th was stationed in a number of places including Algeria, French Morocco, Tunisia, and Italy. Luttrell named his P-38 after his red headed wife from Texas, Billie. Luttrell's A-2 jacket has a beautifully tooled leather squadron patch on the left front panel. The patch is typical of Italian made insignia.

Tooled leather 49th Fighter Squadron patch sewn to left front panel of Luttrell's A 2 jacket. (Michael J. Perry)

Embroidered silk flying scarf and leather patch of the 49th Fighter Squadron. (Michael J. Perry)

Overall front view of Lt. Don A. Luttrell's A-2 jacket. (Michael J. Perry)

Crew chief of "Billie the Red," P-38 Lightning flown by Don Luttrell, holding engine panel with nose art. (Luttrell via Perry)

Right: Lt. Don A. Luttrell in the cockpit of his P-38 Lightning. He is wearing his A-2 jacket. Note that he is also wearing a British Flight helmet, which often was preferred by American pilots. (Luttrell via Perry)

BERT McDOWELL, JR.

Captain Bert McDowell, Jr. flew P-51 Mustangs with the 388th Fighter Squadron, 55th Fighter Group, 8th Air Force. The 55th flew out of Nuthampstead and Wormingford, England and later moved into Germany. McDowell's A-2 is very plain. The main reason for inclusion in this work is the unusual placement of the name strip over the left pocket instead of the conventional breast placement.

Bert McDowell, Jr. in the cockpit of his P-51, "Betty Mac." McDowell is wearing a British model 1944 "C" pattern helmet, which was very popular with American pilots. The goggles are American B-8's and the oxygen mask is an A-14. (McDowell via Perry)

A-2 jacket worn by P-51 pilot, Bert McDowell, Jr., 388th Fighter Squadron, 55th Fighter Group, 8th Air Force. Right: Unusual placement of leather name strip on McDowell's A-2. (Michael J. Perry)

W.E. McGIFFIN

Captain W.E. McGiffin served with the 487th Bomb Squadron, 340th Bomb Group, first assigned to the 9th Air Force, and later the 12th Air Force. The 487th flew B-25 Mitchells and operated from air fields in Egypt, Tunisia, Sicily, Italy, and Corsica. McGiffin's jacket has a nice Italian made, tooled leather 487th Bomb Squadron patch on the left chest under a leather name tag.

Right: A-2 jacket worn by Captain W.E. McGiffin, 487th Bomb Squadron, 340th Bomb Group, 12th Air Force. Left: Italian made, tooled leather squadron patch of the 487th Bomb Squadron on McGiffin's A-2. (JS Industries)

Europe, Africa & the Mediterranean

ALEXANDER J. McNAIR – "WHY SHUR!"

A.J. McNair entered military service on February 12th, 1943 at Fort Sam Houston and graduated from navigator's school in February of the following years, having also qualified as an air gunner. He went to the E.T.O. with the 544th squadron of the 384th Bomb Group and flew 30 missions in B-17s. During his tour in the air, McNair flew as lead navigator for his wing four times, and for his group, seven times. He also participated on the "Grapefruit" mission to Koln, Germany when his squadron used the highly unsuccessful gliding bombs on the target. He chose the name "Why Shur!" for his A-2 jacket because it was his response whenever the crew was complimented for their bombing accuracy! McNair won the Distinguished Flying Cross and the Air Medal with three oak leaf clusters.

Detail of "Texas Sweetheart" pin up girl hand painted on the right chest of A.J. McNair's A-2 jacket.

Detail of the early "short wing" 8th Air Force patch on the left shoulder of A.J. McNair's A-2 jacket. Typically English made, this was the first design of the famous 8th Air Force patch. It was never officially authorized, being replaced by the later pattern with longer upswept wings, but to many it was the only design ever worthy of the title "The Mighty Eighth."

Above right: Front view of A. J. McNair's "Why Shur!" A-2 jacket. Right: Rear view of A. J. NcNair's "Why Shur!" A-2 jacket.

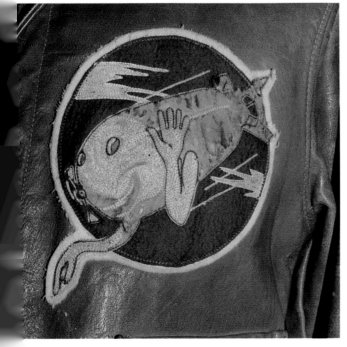

Detail of the unique 544th Bomb Squadron patch on the left chest of A.J. McNair's A-2 jacket. No doubt made somewhere in England, it [is] unusual because of its size and the variety of material used to make it – felt, leather, and silk thread embroidered carefully by machine.

J.R. MANCUSO

J.R. Mancuso served with the 460th Bomb Group, 15th Air Force, on a B-24 Liberator. That is all we can ascertain from the art and insignia on his A-2 jacket. His bomb scoreboard indicates 50 missions. The 460th Bomb Group was made up of the 760th, 761st, 762nd, and 763rd Bomb Squadrons and was stationed at Spinazzola, Italy. The art work is very typical of jackets done in Italy.

Below: Front and rear views of A-2 jacket worn by J.R. Mancuso, 460th Bomb Group, 15th Air Force.(Manion's)

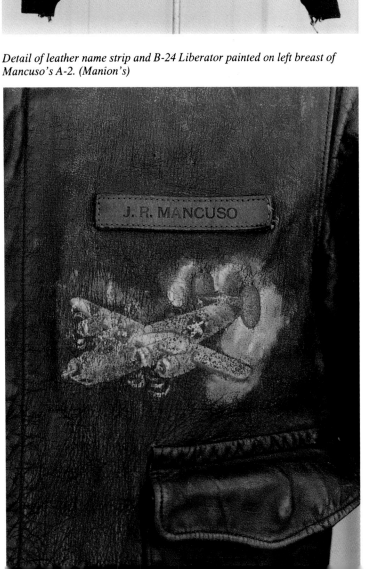

Detail of leather name strip and B-24 Liberator painted on left breast of Mancuso's A-2. (Manion's)

Detail of "Cutie" painted on the reverse of J.R. Mancuso's A-2 jacket. Art work is typical of that seen on Italian painted jackets. (Manion's)

Europe, Africa & the Mediterranean

T.E. MARCHBANKS

Major T.E. Marchbanks served with the 601st Bomb Squadron, 398th Bomb Group, 8th Air Force, out of Nuthampsted, England, and the 155th photo Reconnaissance Squadron, which was assigned to a number of units in England, France, Belgium, Holland, and Germany. His A-2 jacket has leather name tag and Major's leaves, a chenille 601st Bomb Squadron patch, and an embroidered 155th Photo Recon Squadron patch – an interesting combination of insignia!

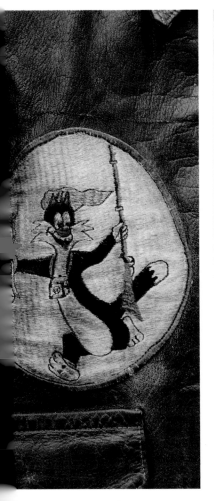

Embroidered 155th Photo Recon Squadron patch on right chest of Marchbanks' A-2. (Manion's)

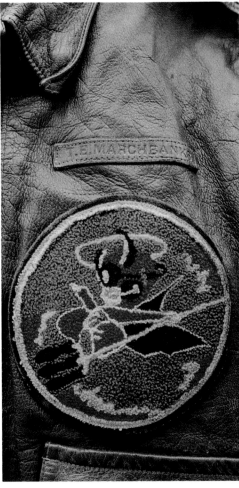

Chenille 601st Bomb Squadron patch and leather name tag on left front of Marchbanks' A-2. (Manion's)

A-2 jacket named to T.E. Marchbanks, 601st Bomb Squadron, 398th Bomb Group, 8th Air Force, and 155th photo Recon Squadron. (Manion's)

Army Air Forces winged star on left shoulder of Marchbanks' A-2. (Manion's)

JACK MASS

The A-2 jacket of Lt. Jack Mass is a testimonial to the tremendous success of the P-47 Thunderbolt used in the fighter/bomber role. "Panzer Dusters" pretty well describes the usual results of a P-47 assault on German panzers. Interestingly, the Republic A-10 Thunderbolt II of Operation "Desert Storm" fame, was named after the P-47, also built by Republic. The P-47 painting markings on Mass' A-2 are those of the 395th Fighter Squadron, 368th Fighter Group, 9th Air Force, late 1944.

Below left: A-2 jacket worn by Lt. Jack Moss, P-47 pilot with the 395th Fighter Squadron, 368th Fighter Group, 9th Air Force. Below right: Reverse of Mass' "Panzer Dusters" A-2. The markings on the P-47 are of the 395th Fighter Squadron. (JS Industries)

Lt. Jack Mass of the 395th Fighter Squadron in the cockpit of his P-47D "Thunderbolt" named "The Down Necker." (Mass via JS Industries)

Detail of exploding tank and "Panzer Dusters" motif on the left chest of Mass' A-2. (JS Industries)

Europe, Africa & the Mediterranean

BRUNO MIAZGOWICZ

T/Sgt. Bruno Miazgowicz flew 26 missions as radio operator/gunner on a B-17 "Second Chance" with the 388th Bomb Group, 8th Air Force, out of Knettishall, England.

Below left: A-2 jacket worn by T/Sgt. Bruno Miazgowicz, 388th Bomb Group, 8th Air Force. Below right: Painting on the back of T/Sgt. Miazgowicz's A-2 featuring a well executed B-17, dice, a bomb with "26," and the name "Second Chance."

N.B. MILSZEWSKI

N.B. Milszewski was a navigator with the 79th Troop Carrier Squadron, 436th Troop Carrier Group, 9th Air Force. Milszewski participated in operations at Normandy, Southern France, and Holland. The 79th primarily flew C-47 Skytrains.

Right: A-2 jacket worn by N.B. Milszewski, navigator with the 79th Troop Carrier Squadron. Above: Detail of leather name strip "N.B. Milszewski" and hand painted navigator wing on left breast of Milszewski's A-2.

"MY PRAYER"

The markings on the B-17G painted on the back of "My Prayer" represent an aircraft of the 834th Bomb Squadron (indicated by the red nose band), 486th Bomb Group (square "W"), 8th Air Force. The complete yellow tail section dates the painting after January of 1945. The falling bombs commemorate 33 missions.

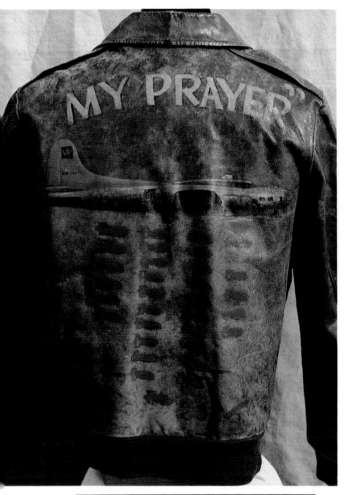

Left: A-2 worn by a crewman on "My Prayer," B-17G, 834th Bomb Squadron, 486th Bomb Group, 8th Air Force. Above: Detail of B-17G "My Prayer" painting. Group and squadron markings are clearly visible. (JS Industries)

CAPTAIN JACK NORTHRIDGE

Jack Northridge was a B-26 pilot for a short time with the 456th Bomb Squadron, 323rd Bomb Group (M), 8th Air Force. He transferred to the 733rd Bomb Squadron, 453rd Bomb Group, 8th Air Force and became a B-24 pilot. His B-24 was nicknamed "Corky." His A-2 jacket dates from his days with the 456th Bomb Squadron.

Left: A-2 jacket and "50 mission crusher" hat worn by Captain Jack Northridge, 456th Bomb Squadron, 323rd Bomb Group, 8th Air Force. The 323rd later transferred to the 9th Air Force. Above: Detail of 456th Bomb Squadron patch sewn to left breast of Northridge's A-2. (Michael J. Perry)

Europe, Africa & the Mediterranean

STEPHEN V. PAINTER, JR.

Stephen Painter graduated from glider pilot training at South Plains Army Air Field, Lubbock, Texas on February 17, 1943. After further stateside training assignments, he went overseas with the 78th squadron of the 435th Troop Carrier Group and participated in the airborne operations into Normandy (D-Day), Holland (Market-Garden) and Wesel, Germany (Operation Varsity). A native Texan, his A-2 jacket is a wearable record of his service career.

Close up of impressed leather name tag and "Lone Star" of Texas insignia hand painted directly on the left chest of Stephen Painter's A-2 jacket. All the painted art work on this jacket was done by Painter himself while he was stationed overseas. (Silent Wings Museum)

Front view of Stephen Painter's A-2 jacket. Note aircraft national insignia visible on CG-4A glider fuselage in background. (Silent Wings Museum)

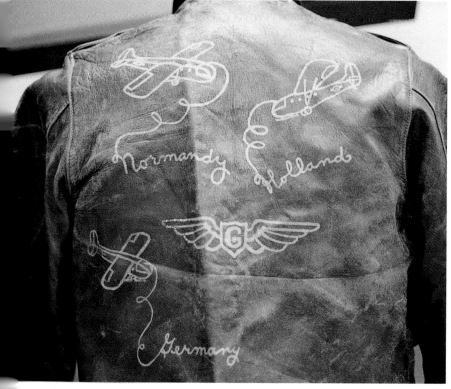

Close up of hand painted campaign credits and likeness of glider pilot wing on the back of Painter's A-2 jacket. (Silent Wings Museum)

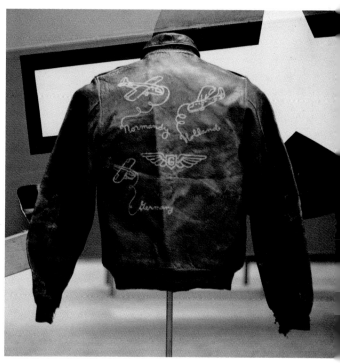

Rear view of Stephen Painter's A-2 jacket. (Silent Wings Museum)

R.L. PARKER

The "Schifless Skonk" was a B-17G of the 568th Bomb Squadron, 390th Bomb Group, 8th Air Force. R.L. Parker was a gunner on the crew. The 568th was stationed at Framlingham, England. Parker was later transferred to the 4th Bomb Squadron, 34th Bomb Group, at Mendelsham, England. The patch on the jacket is that of the 4th, but it is evident that the 568th patch was present. It is interesting to note that "Schifless" is spelled differently on the jacket than on the nose of the B-17. On the jacket, the "C" was left out – no doubt a simple oversight or a different artist. The artwork on the jacket is of very high quality. Also of interest, especially to wing collectors, is the beautifully detailed, English made aircrew wing pinned over the leather name strip.

On 6 March 1944, the "Schifless Skonk" returned from a very tough raid on Berlin. The ship suffered significant battle damage. The dingy hatch behind the cockpit is blown open and the propeller is feathered on the left inboard engine. The top turret gunner, Sgt. Steve Kovacik, is being loaded into the ambulance. The name "Schifless Skonk" is clearly visible on the nose of the airplane. (Jeff Ethell)

Left: Front view of A-2 jacket named to R.L.Parker, 568th Bomb Squadron, 390th Bomb Group, and 4th Bomb Squadron, 34th Bomb Group, 8th Air Force. (Arthur Hayes)

Detail of artwork depicting Adolf Hitler as a "Shifless Skonk." Parker is credited with one German aircraft destroyed, which is represented by the swastika. He also made one parachute jump, as shown in the lower right hand corner (facing) of the jacket. The bomb scoreboard shows 28 missions. Note parachuting airman painting commemorating Parker's jump. (Arthur Hayes)

Left: Rear view of Parker's A-2 jacket. (Arthur Hayes)

Europe, Africa & the Mediterranean

English made aircrew wing, leather name strip reading R.L. Parker, gunner, and 4th Bomb Squadron patch on left breast of Parker's A-2. (Arthur Hayes)

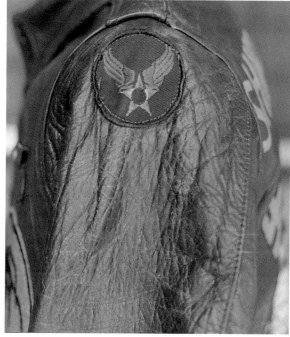

Cotton twill A.A.F. insignia sewn on left shoulder of Parker's jacket. (Arthur Hayes)

Below: A-2 jacket worn by an unnamed U.S. airman from New Jersey, 446th Bomb Group, 8th Air Force. Below right: Close up of painting on 446th Bomb Group A-2 "Patriotic Patty." The painting depicts "Patty" riding a B-24, in the markings of the 446th, superimposed on the U.S. National insignia. The bomb scoreboard indicates 30 missions. (JS Industries)

"PATRIOTIC PATTY"

The airman who wore this A-2 jacket was from New Jersey and flew 30 missions on a B-24 Liberator. The markings on the B-24 in the painting tell us he was assigned to the 446th Bomb Group, 8th Air Force, in mid-1944.

W.E. QUINN

W.E. Quinn was a pilot with the Fifth Photo Group, 12th Air Force. The insignia on his jacket is very typical of that made on the Isle of Capri. The souvenir bell, hooked to the collar, is also a good indicator that the insignia originated there also. We can speculate from the P-38/F5 aircraft on the group insignia that Quinn flew this type of aircraft. The leather wing was made by wetting a piece of thin leather, pressing in a set of silver pilot's wings, and letting it dry, thus leaving the impression. This technique of making wings is seen fairly commonly on 12th and 15th Air Force jackets.

A-2 jacket worn by W.E. Quinn, Fifth Photo Group, 12th Air Force.

Tooled leather Fifth Photo Group insignia (unofficial) on Quinn's A-2. Note two P-38/F5 aircraft and camera in the insignia.

Leather wing and name tag with unidentified squadron insignia on Quinn's A-2.

Leather 12th Air Force patch on left shoulder of Quinn's A-2.

Leather American flag patch on right shoulder of W.E. Quinn's jacket.

Europe, Africa & the Mediterranean

T/Sgt. ROBERT M. RADLEIN

R.M. Radlein was a radio operator/gunner on a B-26 Marauder. Radlein served in the 454th Bomb Squadron, 323rd Bomb Group(M), 9th Air Force. The 454th Squadron was stationed in England, France, Austria, and Germany. Radlein flew 39 missions: two over Belgium, 37 over Germany. Mr. Radlein retired from the U.S.A.F. 30 April, 1966 as a Major.

Left: Rear view of R.M. Radlein's A-2 with painted USAAF insignia; B-26 Marauder of the 454th Bomb Squadron(M), 323rd Bomb Group(M), and 39 mission scoreboard (single bomb with #39). Above: Detail of left front panel of R.M. Radlein's A-2 with leather air gunners wing/name tag and leather 454th Bomb Squadron patch. (Conner)

Robert M. Radlein in 1991 wearing his A-2 jacket. (Conner)

H.A. RUTH

Harold Ruth flew a tour of duty as radio operator/gunner on a B-24 "Flak Hopper," 786th Bomb Squadron, 466th Bomb Group, 8th Air Force, out of Attlebridge, England. Regretfully, the painting is almost completely gone from Ruth's A-2, but with a little imagination, we can still make out a cartoon duck riding a bomb amidst bursts of flak. The name "Flak Hopper" is still fairly clear at the top of the painting.

Crew of B-24 "Flak Hopper," 466th Bomb Group. Harold Ruth is fourth from left, back row. (Ruth)

Right: A-2 jacket "Flak Hopper" worn by Harold Ruth, 786th Bomb Squadron, 466th Bomb Group, 8th Air Force.

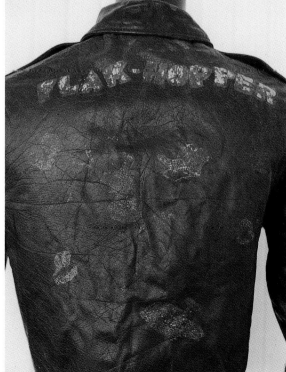

FRED SALEM

"Vicious Veta" was a B-17G in the 728th Bomb Squadron, 452nd Bomb Group, 8th Air Force at Deopham Green, England. Fred Salem was an air gunner on the crew. 30 falling bombs indicate 30 combat missions and the unusual four flour sacks indicate participation in the Berlin Airlift. The paint on Salem's jacket is in beautiful condition.

Right: A-2 jacket worn by Fred Salem, 728th Bomb Squadron, 452nd Bomb Group, 8th Air Force. (Mark Huntzinger)

Reverse of Salem's A-2. (Mark Huntzinger)

Detail of beautifully preserved painting on Fred Salem's jacket. The B-17G "Vicious Veta" is clearly marked with the square "L" of the 452nd Group and the yellow bands would place the aircraft after January 1945. 30 bombs indicate 30 combat missions and the flour sacks commemorate participation in the Berlin Airlift. (Mark Huntzinger)

Right: Detail of leather 728th Bomb Squadron patch on Salem's A-2. (Mark Huntzinger)

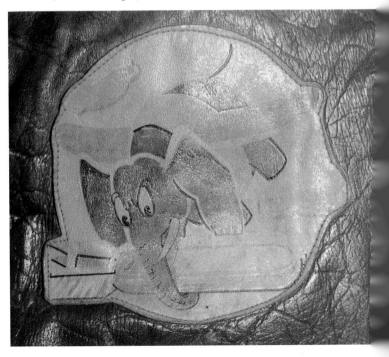

Europe, Africa & the Mediterranean

W.A. SCHULTZ

Captain W. A. Schultz flew a B-17 named "Stricktly Business" with the 419th Bomb Squadron, 301st Bomb Group, 15th Air Force. The 301st was stationed in England, Algeria, Tunisia, and Italy. Schultz's A-2 is very simply done with leather name strip, Captain's bars, and a bullion, early pattern, 15th Air Force patch. Schultz also had an M-41 field jacket with a great painting of a B-17 G in heavy flak on the back.

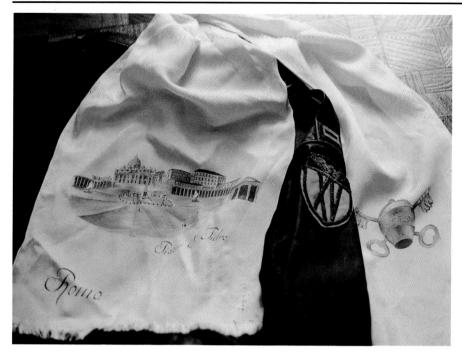

Captain Schultz's A-2 with souvenir flying scarf from Rome. (Michael J. Perry)

A-2 flight jacket worn by Captain W. A. Schultz. (Michael J. Perry)

Left shoulder of Schultz's A-2 with bullion, early style 15th Air Force patch glued in place and leather Captain's bars on the epaulet. (Michael J. Perry)

Detail of painting on M-41 field jacket worn by W. A. Schultz. The B-17 is named "Stricktly Business" and is dropping bombs adorned with the names of various targets. (Michael J. Perry)

Captain W. A. Schultz, 419th Bomb Squadron, 301st Bomb Group, 15th Air Force. (Schultz via Perry)

T.W. SCHWELLBACH

T.W. Schwellbach was a P-47 Thunderbolt pilot with the 366th Fighter Squadron, 358th Fighter Group. The 358th began operations with the 8th, and later moved to the 9th Air Force. The group flew from airfields in England, France, and Germany.

Schwellenbach's A-2. Note stitch marks visible around pockets where Schwellenbach added inside map pockets. Right: Detail of embroidered on twill squadron patch of the 366th Fighter Squadron, and leather name tag sewn to left chest. Also visible is the British "Dingy Whistle" hooked to the throat latch, which was used to signal rescue ships in case the pilot was down in water. (JS Industries)

H.E. SESSLER JR.

H.E. Sessler Jr. was a 15th Air Force navigator. That is about all we know for sure from his A-2 jacket. Typical of 15th Air Force A-2s, the tooled leather insignia was likely made on the Isle of Capri and includes a U.S. flag on the right shoulder. The back of the jacket has an impressive painted list of targets. The bombs beside each target probably indicate the number of mission credits received or the number of trips to each target. There are a total of 57 bombs. Sessler was in theatre from October 1, 1944 to May 9, 1945, as his jacket indicates. Sessler's unit would have flown B-17s or B-24s.

A-2 jacket of H.E. Sessler Jr., a 15th Air Force navigator. (JS Industries)

Europe, Africa & the Mediterranean

Left: Back of Sessler's A-2 with target list and 57 bombs painted on. Above left: Tooled leather American flag on right shoulder of Sessler's A-2. Above right: Worn, tooled leather 15th Air Force patch on left shoulder of Sessler's jacket. (JS Industries)

HOWARD R. SOSSAMON

Howard Sossamon entered the service August 29, 1942 at Camp Robinson, Arkansas. He graduated as a navigator, class of 44-6, AAF Training Detachment of Pan American Airways at Coral Gables, Florida. Sossamon flew 35 sorties (50 mission credits) with the 767th Bomb Squadron, 461st Bomb Group, 15th Air Force, out of Cerignola, Italy. Sossamon's decorations include the DFC (one cluster) and Air Medal (three clusters). Howard Sossamon's A-2 jacket is truly a classic 15th Air Force jacket. It was painted/decorated in a little shop on the Isle of Capri in September of 1944. The tooled patches and hand painted insignia are typical of the work from that area.

Left: Front view of Howard Sossamon's A-2 jacket. Below: Detail of 461st Bomb Group and 767th Bomb Squadron combination insignia in tooled leather sewn to left breast of Sossamon's A-2.

Rear view of Howard Sossamon's A-2 jacket.

Tooled leather American flag sewn on right shoulder of Sossamon's A-2.

Early style 15th Air Force patch with 767th Bomb Squadron tab tooled leather sewn to left shoulder of Howard Sossamon's A-2.

Above: Howard Sossamon (navigator), Joseph Donovan (pilot and C.O.), and Paul Wagner (bombardier) after completion of their last mission. Note Wagner's A-2 has the same insignia as seen on Sossamon's A-2. (Sossamon)

Left: Howard Sossamon wearing his A-2 jacket in front of his tent at Cerignola, Italy. The boy is an Italian named "Franky." Franky would "guard" the tent while the men were on missions and his mother did their laundry. The men would pay her with cigarettes, which she could sell on the black market. (Sossamon)

Europe, Africa & the Mediterranean

"STAR DUST"

Left: Many crippled American bombers were forced to land in Switzerland where the crews were interned. Here two Swiss Air Force officers admire "Star Dust" art work on the back of this American airman's A-2 jacket. (John Campbell)

R.C. TOMLINSON

Captain Tomlinson was a 12th Air Force P-47 pilot. We can guess that he was with the 346th Fighter Squadron, 350th Fighter Group from the black and white checkered rudder marking on the unofficial squadron patch. This was the only 12th Air Force P-47 unit to mark their aircraft in this manner, so it is probably a safe bet. The tooled insignia, embossed leather wing, and painting were most likely done on the Isle of Capri.

Below right: Tooled leather squadron patch (most likely the 346th Fighter Squadron, 350th Fighter Group), 12th Air Force patch, and embossed leather pilot's wing and name tag sewn to Captain R.C. Tomlinson's A-2 jacket. Right: Painting of a diving P-47 Thunderbolt and nickname "Tommy" on the back of Tomlinson's A-2. (Dale Edwards)

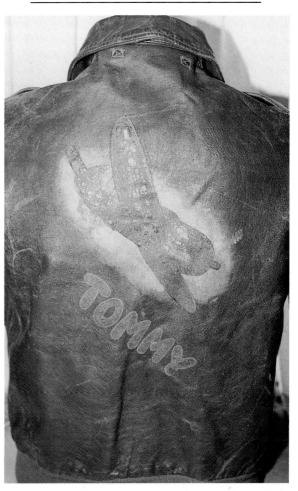

JAMES R. STARNES

Left: Col. James R. Starnes (U.S.A.F., Ret.) is a six victory ace of the 501st Fighter Squadron, 339th Fighter Group. His P-51 Mustang was named "Tarheel." Starnes went on to fly F-4 Phantoms in Vietnam. (Colones)

ROBERT VICKERS

Captain Robert Vickers was a B-24 pilot with the 578th Bomb Squadron, 392nd Bomb Group, 8th Air Force. The 392nd Bomb Group was stationed at Wendling, England. His B-24 was "Dugan." Vickers' A-2 jacket is nicely painted and well preserved. His bomb scoreboard indicates 30 missions.

Above: Right shoulder of Vickers' A-2 with aircraft tail markings of the 392nd Bomb Group, 578th Bomb Squadron painted on. Above left: Left shoulder of Vickers' A-2 with painted 8th Air Force patch. (Michael J. Perry)

Front view of Vickers' A-2 jacket. (Michael J. Perry)

Right front panel of Vickers' A-2 with painted B-24 Liberator and 30 mission scoreboard. (Michael J. Perry)

Left front panel of Vickers' with nickname "Vic" painted on. (Michael J. Perry)

Rear view of Captain Robert Vickers' A-2 jacket. (Michael J. Perry)

Captain Robert Vickers. (Vickers via Perry)

Europe, Africa & the Mediterranean

RICHARD WADE

Richard Wade was a glider pilot with the 35th Troop Carrier Squadron, 64th Troop Carrier Group, 51st Wing, 12th Air Force. Wade's A-2 is typical of 12th Air Force jackets, in that it was painted by a girl on the Isle of Capri. The tooled leather insignia was made there as well.

Tooled leather 12th Air Force patch on the left shoulder of Wade's jacket. (Wade)

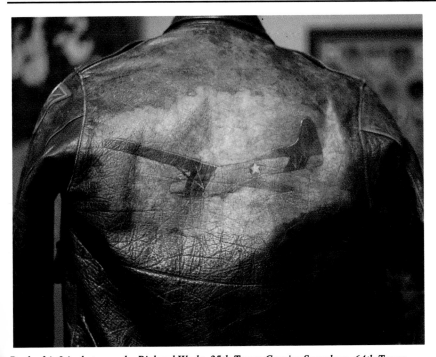

Back of A-2 jacket worn by Richard Wade, 35th Troop Carrier Squadron, 64th Troop Carrier Group, 12th Air Force. (Wade)

Right: Tooled leather name tag and glider pilot's wing removed from Wade's A-2. (Wade)

MICHAEL WHITE

Michael White was a crewman on "Dame Satan II," a B-17G, 322nd Bomb Squadron, 91st Bomb Group, 8th Air Force. The 91st operated from Bassingbourn, England. Michael White's jacket art is a hand painted leather patch depicting "Dame Satan," as on the aircraft nose art.

Above: B-17G "Dame Satan II," 322nd Bomb Squadron, 91st Bomb Group, 8th Air Force, with crew. It is not known if Michael White is in this photo. (NASM via Valant)

Right: Detail of hand painted leather patch depicting "Dame Satan" sewn to left breast of White's A-2. (Charles Stansell)

Front view of Michael White's A-2 jacket. 322nd Bomb Squadron, 91st Bomb Group, 8th Air Force. (Charles Stansell)

"THE WILD HARE"

"The Wild Hare" was a B-17G of the 324th Bomb Squadron, 91st Bomb Group, 8th Air Force. She was lost on a mission 26 November 1944. Records indicate she was a victim of fighters and crashed near Furstenau. Serial number of "The Wild Hare" was 42-31515 and her pilot was Robert J. Flint. This aircraft was from the same squadron and group as the famed "Memphis Belle." It is not known which crew member wore this A-2 jacket. A 1945, 91st Bomb Group calendar, which was acquired with the jacket is inscribed "Remember us Smitty! Tommie "Body" Lehman."

B-17G "The Wild Hare," 324th Bomb Squadron, 91st Bomb Group, 8th Air Force with crew. (NASM via Valant)

Front of "The Wild Hare" A-2 jacket.

Rear of "The Wild Hare" A-2.

Right: Detail of 324th Bomb Squadron patch – hand painted on leather and sewn to "The Wild Hare" A-2 jacket.

Europe, Africa & the Mediterranean

Below: A-2 jacket worn by F.L. Wood, Jr., 597th Bomb Squadron (M), 397th Bomb Group, 9th Air Force. Below right: Back of Wood's A-2. (JS Industries)

F.L. WOOD, Jr.

F.L. Wood, Jr. served with the 597th Bomb Squadron (M), 397th Bomb Group, 9th Air Force. The 597th flew B-26 Marauders from bases in England, France, and Holland.

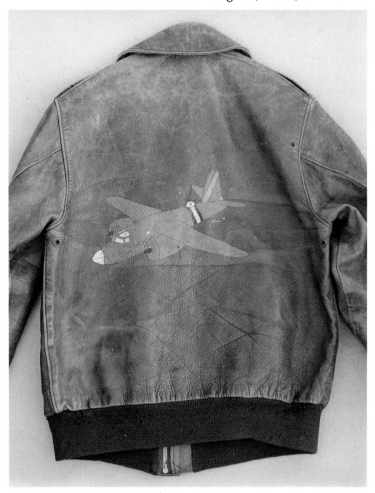

Above: Detail of B-26 Marauder painted on the back of Wood's jacket. The yellow tail strip was the group marking of the 397th Bomb Group. Right: Detail of the 597th Bomb Squadron insignia painted on left chest of Wood's A-2. Placement of the name under the insignia is unusual. (JS Industries)

UNUSUAL "WHITE COLLAR" B-10 JACKET

This unusual specimen of the early B-10 flight jacket sports a white, rather than regulation brown fleece collar. This collar is original to the jacket but it is unknown why this production variant exists. Other examples have been noted in period photos and there seem to be no logical explanation. The Italian made leather squadron patch on the left chest would indicate the jacket had seen combat service in the Mediterranean. Other than the name of the owner via the leather tag on left chest, little else is known.

Close up of the Italian made, tooled and painted, leather Panther head squadron patch sewn to the left chest on the "White Collar" B-10 jacket. Note also the embossed leather name tag and faint factory printed "Army Air Forces" and patch design on left shoulder.

Left: Front view of the odd variation B-10 jacket with white fleece collar.

UNNAMED 8th AIR FORCE

This A-2 has a very simple design of a 30 bomb mission scoreboard superimposed on the 8th Air Force insignia. A record of survival for an 8th Air Force bomber crewman.

Unnamed 8th Air Force A-2 jacket. (JS Industries)

Europe, Africa & the Mediterranean

UNNAMED 15th AIR FORCE

The airman who wore this jacket was a 15th Air Force bombardier with 35 missions on a B-24. The insignia is Italian made tooled leather with the exception of the squadron patch, which is on fabric. The bomb mission tally is unusual in that it is done on a bomb shaped piece of leather and sewn on to the right chest of the jacket.

Left: A-2 jacket worn by an unknown 15th Air Force bombardier. (JS Industries)

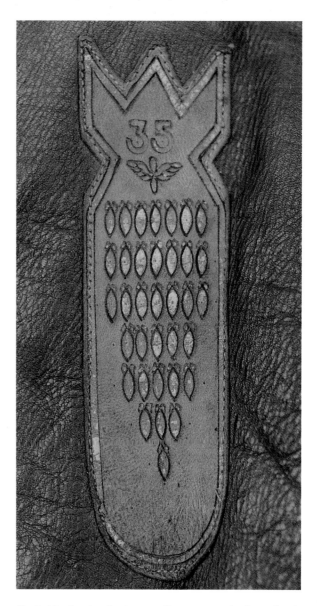

Tooled leather bomb mission tally sewn to right chest of 15th Air Force A-2. (JS Industries)

Left: Tooled leather bombardier's wing and fabric squadron patch, which features a circus monkey dropping bombs from a B-24, sewn on the left chest of 15th Air Force bombardier's jacket. (JS Industries)

UNNAMED E.T.O.

The targets listed on each of the 30 bombs on this A-2 point to the 8th or 15th Air Force. The tooled leather squadron insignia has more of an Italian made 15th Air Force look about it, but until the unit can be accurately identified, it is left to speculation. The name tag has been removed, so now the jacket remains as a record of 30 tough missions in a heavy bomber, by a U.S. airman.

Below: A-2 jacket of an unnamed U.S. airman in the European theatre. Below right: Back of unnamed E.T.O. A-2 jacket with detailed record of 30 missions painted on. (JS Industries)

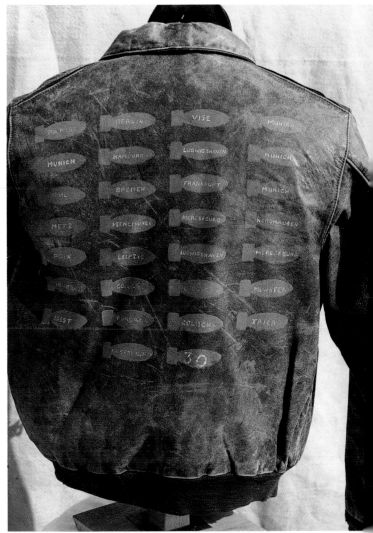

Unidentified squadron patch on left chest of this A-2. The tooled leather construction may indicate that it is Italian made, which would point to the 15th Air Force. (JS Industries)

Europe, Africa & the Mediterranean

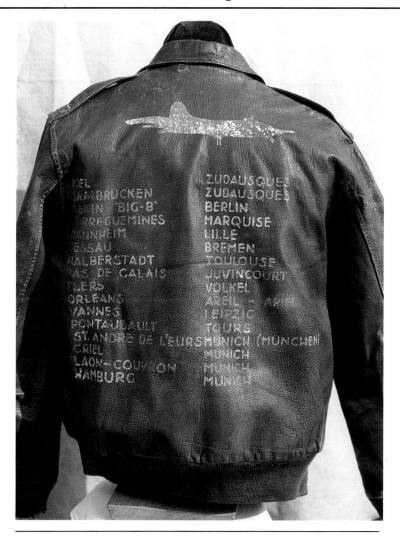

UNNAMED E.T.O.
The U.S. airman who wore this jacket flew 32 missions on a B-17. The target list indicates 8th or 15th Air Force, but that is about all we can ascertain from the information at hand. The jacket is unusual in that it is made of goatskin and it is a large size (44). (JS Industries)

330th BOMB SQUADRON "ETO" JACKET
The ETO jacket represented efforts of both the British and U.S. governments to clothe the G.I. for cool/cold European weather. A spin-off the British issue battle dress jacket, the "ETO" jacket (named for the European Theatre of Operations from which it sprang) was a Lend-Lease product supplied to U.S. forces in England. This example is a unique blend of a service and flying garment.

Front view of the British produced "ETO" jacket. Designed as a cold weather service garment for field jacket liner. This example has been modified for use in flight.

Far left: Close up of insignia sewn to the left chest of the "ETO" jacket. Note hand stitched bullion air crew wing on "Combat Flight Duty" blue background and the large, felt, 330th bomb squadron patch. The left chest pocket was removed for placement of this insignia. Left: Close up of the Eighth Air Force patch and felt composition Sergeant's chevron on the left sleeve of the "ETO" jacket. The companion chevron and an engineer's specialty patch are sewn to the right sleeve.

CHAPTER IV

Unidentified Theatre & Others

With the passing of 50 years or so, many of these jackets have gone through several owners. They turn up in flea markets, garage sales, thrift stores, etc. At any rate, if the jacket has no traceable markings, it is sometimes impossible to uncover any concrete information about its origins. Even under these circumstances, authentic, unidentified jackets are still wonderful pieces, and worthy of inclusion in this work.

This chapter also includes jackets from stateside units, who performed a vital role in training of new airmen and replacements, as well as transportation of materials and equipment, throughout the war. These critical personnel are often forgotten, but their function played a major part in the Allied victory of World War II.

Front view 13th Bomb Group A-2 jacket.

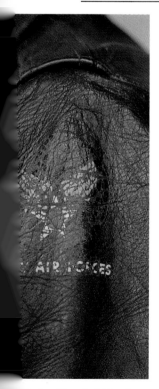

Left shoulder, 13th Bomb Group A-2 with partially remaining Army Air Force decal.

Detail of tooled leather 13th Bomb Group patch.

13TH BOMB GROUP

The 13th Bomb Group was an anti-submarine group which served with the 1st Air Force in the American theatre. It was only in service from 20 November, 1940 until 30 November, 1942. The group flew B-18, B-25, and A-29 aircraft.

OPPOSITE: Yvonne C. Wood, W.A.S.P. (see also page 205)

89TH TROOP CARRIER GROUP

The 89th Troop Carrier Group was a transition training unit for C-47 pilots and later trained replacement crews in the United States. The 89th was assigned to A.T.C. and later to 1st Troop Carrier Command.

Right: A-2 jacket worn by a member of the 89th Troop Carrier Group. (JS Industries)

327TH FIGHTER SQUADRON, 328TH FIGHTER GROUP

The 327th Fighter Squadron, 328th Fighter Group, was an air defense and operational training unit stateside, flying P-39 Airacobras. It later became a replacement training squadron.

Left: Three American flyers wearing A-2 jackets. Two have Chenille 327th Fighter Squadron patches on the left Chest. (Campbell)

FERRY COMMAND M-41 FIELD JACKET

This jacket is actually a wartime, commercial copy of the famous G.I. issue garment. It is exact in almost every detail but it features a commercial label and provision for a separate button in liner. The fact that its original owner was a member of the Army Air Force Ferry Command may indicate that he was a "Contract" (non-military) flyer and had to purchase this jacket outside of military circles.

Right: Front view of the unusual "Private Purchase" M-1941 field jacket.

Left: Close up of the machine embroidered on brown felt ferry command patch and blue felt AAF sleeve patch located on the left side on this commercially procured jacket.

G.S. BUCHANAN
Captain G.S. Buchanan was a test pilot at Wright Field and was permitted to wear the official Wright Field insignia.

Left: A-2 jacket worn by G.S. Buchanan, test pilot, Wright Field. Below: Detail of "G.S. Buchanan" name strip and leather Wright Field patch.

Detail of tooled leather pilot wing, name strip, and canvas squadron patch on Dippy's A-2.

R.N. DIPPY
Very little information is available on R.N. Dippy other than the fact that he was a pilot in the Army Air Force in WWII. The tooled leather pilot wing is similar to those found on 15th Air Force A-2 jackets from Italy, but the canvas squadron patch is more often found on 8th Air Force jackets from England. At this time, the squadron is unknown.

A-2 jacket named to R.N. Dippy.

P.B. GRIFFITH

Lieutenant General P.B. Griffith was a West Point graduate. His A-2 is of exceptionally high quality. The jacket has a nice chenille on leather 1st Observation Squadron patch (which later became the 41st Photographic Reconnaissance Squadron), a leather name tag, and a senior pilot's wing. The epaulets have leather Major's leaves.

Below left: A-2 of Lieutenant General P.B. Griffith. Below right: Senior pilot's wing, name tag, and Chenille on leather 1st Observation or 41st Photographic Reconnaissance Squadron patch. (JS Industries)

"GRUESOME CREWSOME"

"Gruesome Crewsome" was a heavy bomber, as indicted by the crew of nine (likely a B-17 or B-24). Beyond that, not much can be determined about this interesting piece. Thirty bombs form a chain around the painting, which may indicate thirty missions flown by this American airman.

Back of A-2 jacket "Gruesome Crewsome." Note that there are nine crew members and thirty bombs around the painting. (JS Industries)

Unidentified Theatre & Others

M.W. HENRIE

M.W. Henrie was a U.S. Army Air Force bomber pilot on an airplane named "Homeward Angel." The record on the back of his A-2 indicates 32 missions. At the bottom of the back are three half bombs, which may signify scrubbed missions. There also is a nice outline of a pin-up girl. The theatre and type of aircraft are unknown.

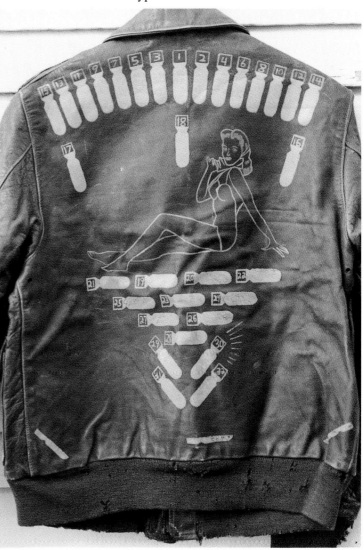

Below: A-2 jacket of M.W. Henrie, pilot, U.S. Army Air Force. His bomber was named "Homeward Angel." Below right: Pin-up girl and 32 mission bomb record on Henrie's A-2. Note three half bombs at the bottom of the jacket. (JS Industries)

E.C. JOHNSON

E.C. Johnson's jacket stands as an amazing record of survival of 65 missions over enemy territory. Although the design is simple, it still shows creativity and balance with the offset name across the back. Unfortunately, there is nothing on the jacket to indicate his unit, theatre of operations, or type of aircraft.

A-2 worn by E.C. Johnson who survived an amazing 65 combat missions. (45th Infantry Division Museum)

"LITTLE LULU"

"Little Lulu" is the subject of both A-2 jackets which follow. As the design and style of the paintings are very different, we can guess that they come from two different airplanes. There is no indication as to theatre of operation or type of aircraft in which the original owners served.

Below left: Unnamed A-2 jacket with "Little Lulu" cartoon motif. Below right: A second example of a completely different design of "Little Lulu" – also unnamed. (JS Industries)

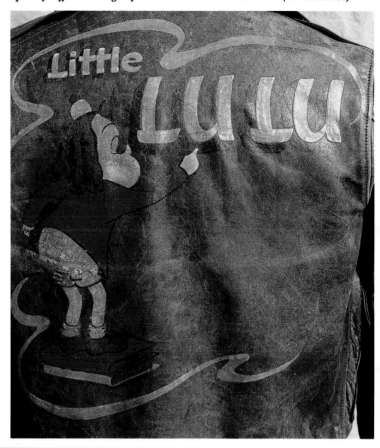

LT. EVELYN R. ORDWAY

Lt. Ordway was a flight nurse, as the wing on her name tag indicates. The patch on her jacket features four ducks flying a stretcher.

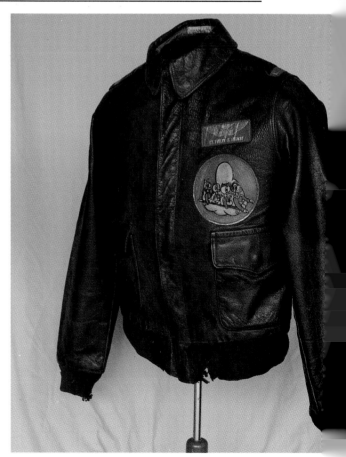

Right: A-2 jacket of Lt. Evelyn R. Ordway, flight nurse. Left: Detail of flight nurse wing/name tag and insignia featuring four ducks piloting a stretcher on left chest of Ordway's A-2.

Unidentified Theatre & Others

J. RIDDLE

Little information is available on Riddle other than his A-2 jacket has the patch of 1st Troop Carrier Command on the left chest. The 1st Troop Carrier Command trained troop carrier organizations and personnel in the American Theatre.

Detail of name tag and 1st Troop Carrier Command patch on Riddle's jacket. (Silent Wings Museum)

Right: A-2 jacket named to J. Riddle, 1st Troop Carrier Command. (Silent Wings Museum)

"SCOTTIE"

"Scottie" served on the crew of "Texas Queen out of Chute 13" and flew 35 missions. That is about all we know for sure. The style and design of the painting is typical of 8th Air Force heavy bomber units, but it could have come out of any theatre of operations.

Left: "Scottie" – an American airman with 35 missions on "Texas Queen out of Chute 13." (via Prodger)

Right: Beautiful painting on the back of "Scottie"'s A-2 jacket. (via Prodger)

H. ROSENBERG

H. Rosenberg flew 40 missions as a bombardier on an airplane named "Blackjack." Unfortunately, that is all we know about this American officer. His jacket has a unique scoreboard on the right front panel with bombs in groups of three raining down. The left front panel has a nice stylized bombardier's wing, and on the back, a faded, but clear "Blackjack" painting.

B-26 Marauder with nose art similar to that on Rosenberg's jacket. It is possible this is the aircraft on which Rosenberg served. (John Campbell)

"Blackjack" A-2 jacket named to H. Rosenberg – an American bombardier with 40 combat missions.

Above: Detail of "Blackjack" painting on back of Rosenberg's A-2.

Rear view of Rosenberg's A-2.

Right: Painted bombardier's wing over name tag "H. Rosenberg."

Unidentified Theatre & Others

Twill A.A.F. patch on right shoulder. (Manion's)

Detail of hand painted Tow Target Squadron patch and leather name tag on left chest. (Manion's)

V.W. TRUSTY

A red rabbit, towing a target soc in the clouds on his patch, tells us V.W. Trusty served with a Tow Target Squadron. There is evidence of another patch having been on the jacket, which may have been his overseas unit before rotating home.

A-2 jacket named to V.W. Trusty. (Manion's)

Front view of Yvonne C. Wood's A-2 jacket. Relatively small size (men's 36) is evident in this photo.

Right: Close up of the famous WASP "Fifinella" insignia designed as a mascot for the Women's Army Service Pilots by Walt Disney. This example has been color printed on a thin piece of undyed leather, then sewn directly to the left chest.

YVONNE C. WOOD – W.A.S.P.

Yvonne Wood was one of the few Women's Army Service Pilots of World War II. Accepted into military training as civilians, the WASP's performed many stateside flying duties, which freed male pilots for combat or overseas duties. Yvonne Wood completed her flight training with Class 43-7 at Sweetwater, Texas. She served as a staff pilot for the Third Weather Region at Kelly Field, Texas, a tow target pilot at Moore Field, Texas, and as a staff pilot at the Army Air Force Navigation School at San Marcos Army Air Base in Texas. Although photos do not readily reflect, Wood's Army Air Force issue A-2 jacket was modified to make it fit the female form just a little better than it was designed to. The "Fifinella" insignia sewn to the left chest was designed for this elite group of fliers by Walt Disney Studios. (See also page 196)

"WYOMING"

The airman who wore this A-2 jacket hailed from the great state of Wyoming and flew 30 missions over hostile territory.

Right: Rear view of A-2 jacket "Wyoming." Below: Detail of painted "Wyoming" and 30 mission bomb scoreboard. (Michael J. Perry)

UNNAMED B-15

This B-15 jacket has a clever cartoon of an officer duck describing his kill to his crew "Der Im Wuz" (There I was). The paint is well preserved. The author has seen an A-2 jacket with the same cartoon and it appeared to have been painted by the same artist, as well.

Right: Front and rear of unnamed B-15 jacket "Der Im Wuz."

Unidentified Theatre & Others

UNNAMED AIR RESCUE SERVICE

This B-15 B jacket was worn by a U.S. airman in the Air Rescue Service, Labrador, Newfoundland, Greenland area.

Left: B-15 B jacket, Air Rescue Service. Below: Detail of beautifully made, multi-piece, cloth and embroidered A.R.S. patch on left chest of unnamed B-15 B. (JS Industries)

UNNAMED "BLACK WIDOW"

This painting on the back of an A-2 features three different versions of the black widow. In the lower left corner is the spider, in the upper right corner is the airplane (P-61), and in the center is the scantily clad female version, all superimposed on a white shield with two black stripes. The individual who wore this jacket was obviously a crewman on a P-61 "Black Widow" Night Fighter. The theatre of operations and unit are unknown.

CHAPTER V

Navy & Marine Corps Jackets

It is a credit to the initiative and fortitude of the men of Naval aviation that their role and method of operation be so uniquely different and yet so completely independent of their Army and later Air Force counterparts. The flight jackets they have worn have been no exception to the many contrasts visible between the aviators of these service branches. The mainstay leather flight jacket of Naval aviation has undergone three different designations, G-1 being the last, but has remained basically unchanged from its inception in the 1920s. It's interesting to note that one of the designation changes was an attempt to introduce it to Army inventories late in WWII. The G-1 jacket remains today as the working "badge" of a naval airman.

It's important to note that airmen of the Marine Corps receive their training under Navy direction, within naval programs and are considered Naval Aviators/crewman, regardless of their allegiance to the Marine Corps. There were no separate flying garments issued to Marine Corps personnel so there will be no special category for them outside of specialized insignia that was applied to denote specific unit affiliation of individual ownership.

The Navy has always maintained a distinctly separate inventory of flying garments, even if the only difference was as minimal as the shade of green they were dyed. The Navy has also been distinctive from other service branches in that it has required a greater number of lightweight or intermediate jackets. One obvious reason for this was the task of providing direct support to the ships of the fleet, a job that did not require a lot of high altitude flying. The counter part of this was, of course, inflicting damage on ships of the enemy's fleet, again not a high altitude task until well into pressurized and heated cockpit years.

Specialized mission requirements generated several variations of unlined cloth flight jackets over the years. This was also due to the relatively minimal opposition by enemy forces on the Atlantic Ocean, thus confining most of combat naval aviation operations to the warm climate of the Pacific. In most regards the flight jackets of the Navy have followed the same paths of transition as other service forces. Influenced by practicality, comfort, availability of materials and economics, the chain began with leather, fur, and cotton and progressed to hybrids of flame proof cloth and insulated nylons. Unlike other services at time of conflict, Naval airmen seem to have enjoyed the least amount of freedom in the personalization of their flight jackets. This was due, in part, to close supervision of personnel while shipboard. Another reason was again warm climate operations which did not generate a lot of interest in the appearance of a jacket that was rarely, if ever, worn.

M421A

This is an unusual version of the USN summer flying jacket, type M421A. It has never been adorned with insignia but is odd in the sense that it has 3 rather than 2 pockets. The reason for variety in this regard is unknown. The cloth flight jacket was very popular with Navy pilots during the war.

Front view of the WWII USN M421A summer flight jacket. It is constructed entirely of tan cotton and is an unusual 3 pocket version. Most were made with only the 2 lower pockets.

OPPOSITE: Marine pilot, July 1943. (Sullivan via Campbell)

JOHN C. DOHERTY

The Navy issue M-421A summer flight jacket is not often found with an abundance of insignia. This was probably due to its limited usage and washability because of its cotton construction. This example has several insignia and does show some fatigue due to laundering. There are some interesting authorized and unofficial insignia on it and the leather name tag indicates it was the property of a fairly high ranking officer.

Front view of the John C. Doherty US Navy M-421A flight jacket.

Close up of the embroidered felt squadron patch sewn to the right chest of the Doherty M-421A jacket. The design seems to indicate the squadron functioned in a support rather than combat role.

Detail of the left front interior of the Doherty M-421A jacket. The printed rayon "Blood Chit" has been sewn into the inside of the jacket and is the issue type common to air crew operating in the China-Burma-India theatre. The flap located to the right of the chit is actually the body of an interior pocket in the left chest area. It was stenciled with the letters "USN" during production.

Detail of the unusual novelty patch sewn to right shoulder of the Doherty M-421A jacket.

Left: Gold leaf impressed leather name tag sewn to left chest of the Doherty M-421A jacket. The light brown color is typical, but not the rule, for WWII examples. Note that the Commander rank title has been spelled out completely!

Locally produced, printed cotton China-Burma-India area command patch sewn to left shoulder of the Doherty M-421A jacket. The peculiarities of foreign produced insignia always stand out and function as visual proof that the wearer was "There." Note also the black leather pencil pocket added by the owner.

Navy & Marine Corps Jackets

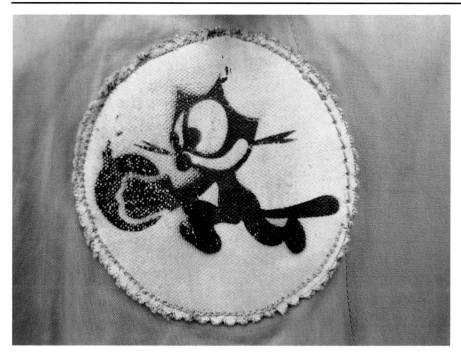

Detail of the unusual VF-6 patch sewn to the right chest of the USN M421A jacket. Made of paint impregnated canvas, this patch is typical of WWII Navy patches produced aboard ship.

VF-6

This example of the tan cotton USN M421A summer flight jacket is the standard pattern-zipper front with two lower patch pockets, but it has had an unusual VF-6 patch sewn to the right chest. Made of paint impregnated canvas, this patch is typical of many WWII vintage Navy types produced aboard ship during the war.

Right: Front view of the World War II USN VF-6 M421A Summer Flight Jacket.

Detail of the gold leaf impressed leather name tag sewn to left chest of the USN M-716 summer flight jacket.

J.A. SCOTT

This tan cotton jacket bears the USN Buaero designation type M-716 on its woven nomenclature label. Although similar to the M-421A in many regards, this example has 2 chest pockets in addition to the 2 standard lower pockets.

Front view of the WWII USN M-716 summer flight jacket.

WWII USMC 1941 FIELD JACKET- VMR 953/MAG 15

This jacket is a standard issue, Model 1941 field jacket as issued to the armed forces in WWII. This jacket was a highly desirable garment to G.I.s due to its comfort and plain good looks. The markings and art work can be classified as unique due to the fact that they are indeed hand done, directly on the fabric with India ink and some type of highly controllable stylus. Examples of this type were not uncommon but survivors are very limited due to the vulnerability of this light weight cloth jacket.

Rear view of the USMC Aviation M-1941 field jacket with hand drawn "War Art."

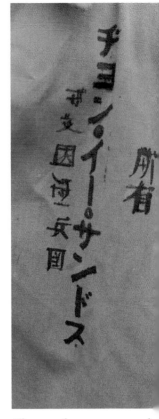

Close up of owners name and service affiliation hand inked oriental characters on left chest.

Front view of an unusual WWII issue M-1941 field jacket that was issued to an enlisted combat aircrewman with VMR-953.

Right: Close up of the central design found hand drawn in India ink on the reverse of the WWII M-1941 field jacket. Note the near life size rendition of the Navy-Marine Combat air crew wing directly above this comical rendition of the "ruptured duck" discharge insignia.

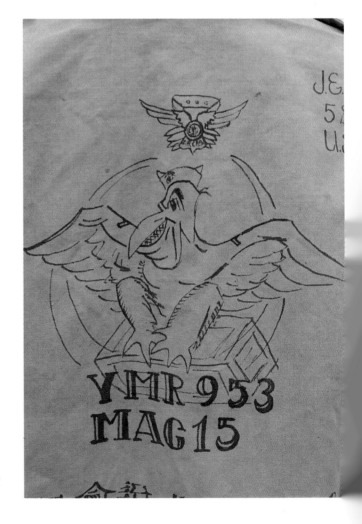

Navy & Marine Corps Jackets

Left: Front view of WWII U.S.N. issue G-1 leather flight jacket. (Charlie Wilhite)

Right: Close up of the machine embroidered torpedo squadron patch sewn to right chest of the G-1 jacket. The "Big Bad Wolf" clutches a full size torpedo with an imposing blue-green devil head in the background. A well worn, gold leafed leather USN aviator wing name tag bearing the name "B.L. Arnold" is still sewn to the left chest of this jacket.

Right: Close-up of the collar-specification tag area of the torpedo Squadron G-1 flight jacket.

B.L. ARNOLD

This WWII U.S.N. issue G-1 leather flight jacket is like so many others in the sense that its original owner has long since been separated from it and all that remains are clues for the investigative collector or historian. Complete with a legible name tag and a squadron patch that clearly reflect the unit's role, the jacket is an exciting relic of naval aviation torpedo operations.

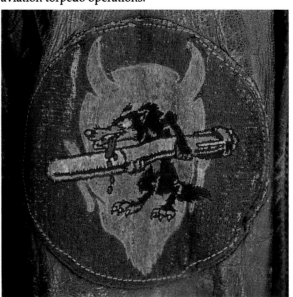

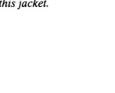

Left: Front view of Tech. Sgt. A.J. Bibbee's G-1 leather flight jacket. Note button tab collar closure strap visible on the lower left side of the collar.

Right: Gold leaf embossed tan leather name tag sewn to left chest of T. Sgt. Bibee's G-1 jacket. Apparently Sgt. Bibee was a non-commissioned pilot. This status of aviator comprised a very small group during the war!

TECHNICAL SERGEANT A.J. BIBEE-USMC

This medium brown shade, goatskin constructed WWII issue G-1 leather flight jacket has a very full, lamb's wool collar that extends downward in a very pronounced point on the lower edge of each side. It shows little wear or age which is not uncommon. Many of the G-1 leather flight jackets issued to Navy and Marine aircrewman during the war attained an instant "souvenir" status due to the mostly hot climate operations naval air elements functioned in during the war. A great number of these jackets were probably never worn in combat throughout the service time of their owners. This example is somewhat unique owing to the fact that its original owner was apparently a non-commissioned pilot.

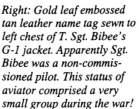

"BLACK CATS"

The history of naval aviation has many pages filled with the exploits of PB-Y aircraft and the men who flew them, but few stories equal the legendary status of the "Black Cats." Their daring night operations against the Japanese during World War II are second only to the reputation of the unique aircraft they flew. Like many of the World War II flight jackets now in the hands of collectors and historians, this jacket lacks background information concerning its original owner, but the presence of an Australian hand made "Black Cats" squadron patch leaves no room for doubt that this old jacket was indeed "there."

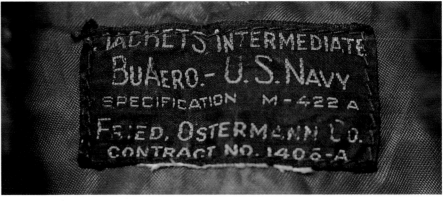

Above: Close up of the U.S. Navy label inside the "Black Cats" G-1 jacket. Note this label carries the original U.S. Navy designation "M-422A" which pre-dates the G-1 designation. G-1 has been used throughout the book to maintain uniformity and eliminate confusion in identification, since there is no significant difference between them.

Left: Front view of the WWII U.S. Navy "Black Cats" G.I. jacket. Note the full, lamb's wool collar and the red satin lining that were both typical of the wartime issue G-1 jacket.

Right: Close up of the Australian hand embroidered squadron patch sewn to the right chest of the "Black Cats" G-1 jacket. This squadron is known to have had more than one variation of its patch design during the war but all variants bear the "Black Cats" caption.

V.G. HOWARD

As an element of the U.S. Naval Aviation Program, the Marine Corps was issued Navy flight gear. The G-1 jacket was often a part of the issue to officers and enlisted men as well. Little is known about this example, but it seems to have been worn by someone who did use it owing its condition and conformity to only what was necessary in the form of insignia. The placement of the Marine Corps insignia instead of an aviation badge might indicate the owner was not actually an aircrewman.

Large stencil painted Navy issue markings on the underside are typical of wartime issue G-1 jackets as seen in this photo of the jacket used by Master Sergeant V.G. Howard, U.S. Marine Corps.

Right: Front view of the G-1 jacket worn by USMC Master Sergeant V.G. Howard. Note gold leaf impressed leather name tag sewn to left chest of the V.G. Howard G-1 jacket. The Marine Corps device in place of an aviation badge is unusual and it may indicate the owner was not an aircrewman. The inclusion of rank, "Mt. Sgt." and "U.S.M.C." are also interesting.

W.O. DEATHERAGE

This unique WWII vintage G-1 leather flight jacket was issued to a Coast Guard air crewman active in the "International Ice Patrol" as reflected by the patch still sewn to the right chest. The Coast Guard's very limited role in aviation makes examples of flight jackets from this service very unusual.

Close up of the "U.S.N." stenciled marking on the underside of the W.O. Deaterage G-1 jacket. This marking is typical of all WWII vintage G-1 jackets. (D. Orear)

Above: Front view of the W.O. Deatherage G-1 jacket. Far left: Close up of the unusual Coast Guard air detachment-International Ice Patrol patch sewn to left chest of the Deatherage G-1 jacket. Left: Close of the gold leaf impressed name tag on the left chest of the Deatherage G-1 jacket. The winged propeller in place of an aviator badge would indicate the owner held no formal aeronautical rating. (D. Orear)

C.M. PAJAK

U.S. Navy contract garments were not unlike virtually anything else in production for the armed forces during the war. Due to the large number needed many contractors were employed to meet quotas dictated by "Uncle Sam." This gave way to many variations in things that were for all intents "uniform" in issue. This G-1 jacket is constructed of goatskin as per naval specifications but it is unusually dark for a war-time produced example. It is also interesting to note the very pale brown lamb's wool collar. In almost every sense this jacket retains a near unissued appearance with the exception of its very basic name tag and pristine V.F. 191 squadron patch.

Left: Front view of C.M. Pajak's U.S. Navy G-1 jacket. Far left: Close up of insignia sewn to the left chest of C.M. Pajak's G-1 jacket. The simplicity of the name tags inscription seems to imply the owner had no intention of changing it should any part of his military status change. The squadron patch is U.S. machine produced and is a pristine example of the insignia worn by VF-191 "Satan's Kittens."

TED KRAWCZYK

This U.S. Navy G-1 leather flying jacket came into a collector's possession directly from the original veteran and was accompanied by a wide variety of artifacts as well as some personal narrative on the life and times of a Navy Air Gunner in Torpedo Squadron 87 in WWII. This is a beautiful example of a personalized Navy G-1 leather flight jacket and makes an interesting display when exhibited as an element of one individual's wartime experiences.

Detail of large Japanese characters hand painted on the back of the Krawczyk G-1 jacket. The literal translation of the three characters is: Sea-Sky-Military which can be interpreted as "Naval Air Military Force."

Above and below: Front and rear views of Ted Krawczyk's G-1 jacket.

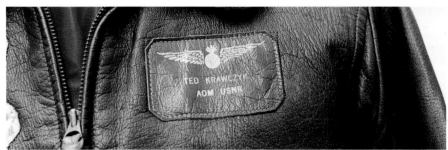

Close up of gold leaf impressed leather name tag on left chest of the Krawczyk G-1 jacket. The winged bomb design is the specialty insignia for a Navy aviation ordnanceman which is indicated below it by the letters "AOM." The clipped corners of the leather tag give it a further personalized appearance.

Right: Close up of the painted canvas VT-87 squadron patch sewn to right chest of the Krawczyk G-1 jacket. Note also the unit's designation painted boldly below.

Right: Close up of an original painted canvas VT-87 patch, like the one of Krawczyk's jacket, which has never been trimmed out and applied to jacket. This is one of 6 extra patches that came in the group of material with Krawczyk's G-1 jacket.

Navy & Marine Corps Jackets

R.A. LUDWIG

This WWII issue U.S.N. G-1 Leather flight jacket is unusual in many respects. Its clean, almost unworn condition is typical of many due to the warm climate theatres of operation in which the Navy was so often active. The short, angular fur collar is of note but most outstanding are the applied insignia. The Army China, Burma, India Command patch has been applied to the left shoulder and the unusual decal on canvas squadron insignia seems to support the China theatre with the Dalmatian walking through a Japanese landscape emblazoned with Mount Fuji in the background.

Left: Front and front quarter view of R.A. Ludwig's G-1 jacket showing the Army C.B.I. Command patch sewn to left shoulder. This was not uncommon for naval forces operating in this theatre. It is thought to have been an accepted practice due to the fact that so much pro-American propaganda had utilized the patch design as an insignia worn by friendly, allied fighters.

Right: Close up of the bold "Tall Dogs" squadron insignia on the right chest of R.A. Ludwig's G-1 jacket. The patch itself is most unusual; it is actually an American produced decal that has been applied to impregnated canvas base and stitched to the jacket. Below: Gold leaf impressed leather name tag sewn to the left chest of the R.A. Ludwig G-1 jacket. Note the very early style aviator wing.

"DICK RICH"

This is an example of the G-1 jacket that reflects the career assignments of a U.S. Navy fighter pilot. It is a late production G-1 with a very dark lamb's wool collar. The application of multiple squadron patches seems to be more typical of the Navy than other branches of the service. Like many others, this jacket and its history have long since been separated from its original owner, but the presence of a readable name tag and all of the original insignia can give an interested researcher some good leads to pursue. It is not uncommon for a motivated enthusiast to "track down" an original owner for the sake of verification and historical background.

Dick Rich G-1 jacket. The name tag indicates Rich was not only a naval aviator, but also held the rank of Commander. Visible in this photo are the patches of VF-121 (upper left), VF-114 (left), VF-82 (right), and VF-92 (far right). "VF" is the designation for a naval fighter squadron. The number of units Rich was affiliated with indicates a distinguished career in a fairly risky line of work. (JS Industries)

HAROLD SPITZBURG

This most interesting example of a wartime U.S. Navy issue G-1 jacket that was worn and personalized by someone who had absolutely no business with it. Harold Spitzburg was a Pacific Island based U.S. Navy seaman 1st class with the specialty of painting aircraft after repair or assembly. Somehow seaman Spitzburg acquired the beautiful G-1 jacket and arranged for the production and application of his unique name tag. Being the only piece of insignia on the jacket, it is none-the-less unique in the fact that it incorporates a naval aviator's wing and an improvised rating designation. Had all this not been explained to the author by Seaman Spitzburg himself, much of the story would forever be one of those annoying mysteries so often found in this field of study and interest.

Right: Front view of the Harold Spitzburg's WWII U.S. Navy issue G-1 jacket. Note its outstanding condition. This was probably due to more than lack of necessity in the hot Pacific climate. Spitzburg was an aircraft painter and would have never rated issue of a flying garment. A valuable commodity in many regards, flight jackets were often the subject of barter for extremely high stakes. It would seem that Spitzberg was successful in his attempt at one. Below: Close up of the unique, gold leaf impressed, leather name tag on the Harold Spitzburg G-1 jacket. Note the extremely small USN aviator wing above his name and the unique rating abbreviation: "PTR" = painter; "V" = Aircraft; "1/C" = First class and, of course, "U.S.N." = United States Navy.

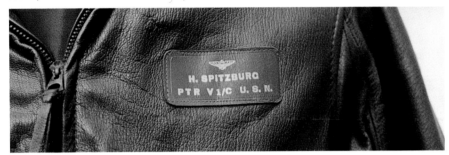

WWII USN G-1 JACKET

VF-6

This WWII issue USN G-1 jacket is dark brown with a rich, full reddish-brown lamb's wool collar. It bears an interesting ship-board produced version of the VF-6 patch on the right chest.

Detail of the VF-6 patch sewn to right chest of the USN G-1 jacket. This is a ship board produced example patch-stenciled basic design on a section of painted impregnated canvas.

Detail of the gold leaf impressed leather name tag sewn to left chest of the USN VF-6 G-1 jacket.

Left: Front view of the WWII USN VF-6 G-1 jacket.

Navy & Marine Corps Jackets

ATKRON 82

This G-1 jacket gives certain indications of being a WWII vintage jacket but the label and markings indicate it is of the late 1950s or early 1960s period. The real fur collar is unusually full and light colored which is uncommon for jackets of this era. Well kept, this jacket was found in an isolated vintage clothing store and carefully cleaned to its present state. It bears no indication of ever having had a name tag on the left chest.

Above: Detail of the large, Asian made Atkron 82 patch sewn to right chest of the G-1 jacket. Below: Detail of the machine embroidered United States flag patch sewn to left shoulder of the Atkron 82 G-1 jacket. (Jim Mirick)

Right: Front view of the ATKRON 82 G-1 jacket. (Jim Mirick)

CORPORAL MONTE DUNDEE, USMC

This G-1 leather flight jacket is the standard USN issue of the mid to late 1960s. Very dark in color, it has a simulated fur collar, a dark brown satin lining and is marked "U.S.N." in punch holes on the leather zipper guard flap in front. It is unusual in the sense that it is adorned with several insignia. Flamboyance was not something Marine Corps enlisted men were known for at that point in time. This jacket symbolizes the modern helicopters' link with the open cockpit days of Marine Corps aviation.

Close up of the HMH-363 Marine helicopter squadron patch on the right chest of Corporal Monte Dundee's G-1 leather flight jacket.

Close up of the unofficial Marine Corps patch located on the right sleeve of Corporal Dundee's jacket.

Front view of a 1960's vintage USN G-1 leather flight jacket with Marine Corps enlisted man's name tag and related insignia.

Close up of left chest and sleeve area showing U.S. flag patch sewn to left shoulder and the worn, gold leaf embossed air crew name tag bearing the Marine Corps insignia over original owner's name and rank.

"PATRON 16-FEARLESS FOUR"

This G-1 jacket is an unusual specimen of the Vietnam era due to its unauthorized oriental made insignia. Showing little wear or age, this jacket was a fairly early contract of the post war era due to its real fur collar. No real clues about the former owner are evident but he was apparently a crewman on the P-3 Orion aircraft of Patrol Squadron 16 during the WESPAC cruise of 1966-67 and had a real fondness for flamboyant insignia! It is doubtful this jacket was ever worn in the line of duty due to its outstanding condition and strong non-regulation appearance.

Below left: Frontal view of the Patrol Squadron 16 G-1 leather flight jacket. Note the U.S. and S. Vietnam flag patches on shoulders plus very colorful VP-16 "Fearless Four" patch on the left chest. The "Tonkin Gulf Yacht Club" was just a novel way for servicemen to say "I've been there."

Below right: Close up of the VP-16 "Fearless Four" patch sewn to the left chest. Oriental machine embroidered construction is evident. Also of interest are the unusual colors.

Below left: Close up of the "Tonkin Gulf Yacht Club" patch sewn to the right chest. This does not appear to have been made by the same source as the rest of the patches on this jacket and may have been made in Vietnam. The slightly irregular oriental machine work is again evident, but this patch was embroidered on a cotton twill background which hastened production time and cost.

Above right: Large WESPAC cruise patch on the back really tells a story. National emblems across the top would indicate contact or function with forces of the United States, China, Japan and the Philippines (from left to right). Most unusual for many reasons, this patch depicts the aircraft and area operation in reasonably good detail. The Philippine insignia, colors used, and style of embroidery strongly suggest this patch was made in the Philippines. This was not uncommon during the Vietnam conflict and it is typical of other insignia confirmed to be of Philippine manufacture.

Navy & Marine Corps Jackets

"HARRAH HAS BETTER IDEAS"

This is an unusual example of a boldly adorned USMC issue G-1 jacket. A very late example of a G-1 with a real fur collar. It is a dark shade and has the dark brown nylon lining typical of examples produced well after World War II. Not only does it bear an unusually plentiful degree of insignia, it is also representative of a small group of Marine Corps airmen – Chinook crews. This jacket shows wear and its faded name tag is impressed "Snoopy Harrah," which coincides with the flamboyant painted motif on the back. Details beyond the obvious are unknown, but a conversation with its original owner would probably yield volumes of information on the subject of Marine Corps CH-46 operations.

Above: Front view of the rather vividly adorned "Harrah Has Better Ideas" G-1 jacket. Note the WWII style Marine Air Wing patches on sleeves and the odd Guidon shaped patch with Marine Corps emblem on the right shoulder. This is a very late example of a G-1 with a reddish-brown, real fur collar. Right: Rear view of the "Harrah Has Better Ideas" G-1 jacket. The motto, likeness of the CH-46 and nickname "Snoopy" are painted directly on the leather – all other visible insignia has been sewn on. (JS Industries)

Above left: Front view of Gilbert F. Hayes' USN G-1 jacket, with large U.S. Navy P3A Orion patch on right sleeve. Note the "Paris '65" tab added below which denotes participation in the Paris Air Show in 1965. Above right: Gold leaf impressed name tag sewn to left chest of the Gilbert F. Hayes G-1 jacket. Note the early stylized aviator wing and the inclusion of rank.

GILBERT F. HAYES

This is an example of one of the last U.S.N. issue G-1 jackets to be produced with a real lamb's wool collar. Of interest in this regard is the reduced size and rounded shape of the collar in comparison to earlier models. Gilbert Hayes graduated from Naval flight school in the early 60's and flew coastal patrol in Orion P-3A aircraft with Patrol Squadron 49 as a member of Crew Two –"Those Magnificent Men and Their Flying Machines." Insignia and general appearance of this jacket is rather unassuming which is typical of the period and duty assignment. It's interesting to note the "Paris '65" tab on the right sleeve. This indicates Hayes and/or his crew had some participation in the Paris Air Show in 1965.

"HEAVY ATTACK SQUADRON 8"

This G-1 jacket is a nice example of an early post-World War II vintage specimen that has been adorned with patches that reflect the owner's long career in naval aviation. Jackets of this type are an interesting reflection of the fact that, in the jet age, the old leather flight jacket is more of a memento than the necessity it was in the open cockpit days of naval aviation.

Below left: Front view of the "Heavy Attack Squadron 8" G-1 jacket. Note the real lamb's wool collar, Task Force 77 patch and the Heavy Attack Squadron 8 patch for which this jacket has been cataloged. Below right: Rear view of the "Heavy Attack Squadron 8" G-1 jacket. There are 4 large patches sewn to this area including an unofficial "Tonkin Gulf Yacht Club" design which denotes operations in Vietnam.

"PHANTOM PHLYER"

The McDonnell Douglas F-4 "Phantom" II jet fighter served the U.S. Navy for many years and was loved and revered by those who flew them. This G-1 jacket is a great testimony to the successful combination of the Phantom and Naval Aviators. The insignia is hand painted and the design is composed of five separate pieces of leather. Nothing is known of its origin or the career of its original owner.

Rear view of the "Phantom Phlyer" G-1 jacket. The central design on the artwork is hand painted on a large piece of leather that has been trimmed to the shape of a typical Navy squadron patch. The outline of the plane has been created by adding four pieces of reddish-brown leather around the edge. (JS Industries)

Navy & Marine Corps Jackets

SUBMARINE PATROL SQUADRON

This is an extremely unusual 1950's era USN issue G-1 jacket that was produced with a natural color, off-white lamb's wool collar. Close examination reveals that this collar is original to the jacket and it is unknown why such a production variant would exist. The large squadron patch sewn to the left chest indicates the unit's role in a anti-submarine capacity. This jacket was not acquired from its original owner so further information is not available.

Below right: Front view of the "Submarine Patrol Squadron" Navy G-1 jacket. The unusual natural shade lamb's wool collar is original to this jacket but it is unknown why such a noticeable variation from the dark brown collar version would exist. Below left: Close up of the novel squadron patch sewn to the left chest on the "Submarine Patrol Squadron" Navy G-1 jacket.

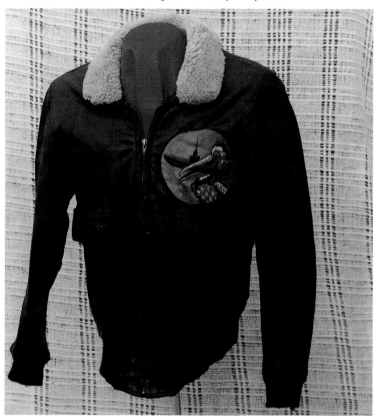

"TORPCATS"

This tired old G-1 was rescued from a dusty closet in an estate sale several years ago. It shows its years of age but still emulates the character of naval carrier based torpedo squadron operations during the war. It is most unusual in the sense that it carries two different torpedo unit patches of very diverse methods of construction. The finish shows fading and wear but the goatskin leather is still strong and pliable. The lamb's wool collar is of a much lighter shade than those to follow in later years of the G-1 jacket's production.

Close up of the distinctive unit patch sewn to right chest of the "Torpcats" G-1 jacket. This is a wartime produced patch made in the United States. A simple machine embroidered cotton twill design, this patch has been widely reproduced in later years, apparently for the collectors market.

Left: Front view of the "Torpcats" U.S.N. G-1 jacket.

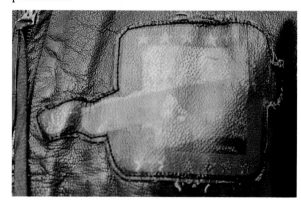

Close up of the hand painted, impregnated Torpedo squadron insignia sewn to the left chest of the "Torpcats" G-1 jacket. Although it has faded and is difficult to see, the design is that of a torpedo in flight with boxing gloves on its forward set arms. The blue sky and ocean are visible in the background.

VP 1 – #38

This issue G-1 flight jacket bears the very simple, stenciled markings of Patrol Squadron One across its lower back. After the simple "VP 1" unit mark it is marked with a large "38" which was probably the personal equipment identity number. This jacket, with its stark and practical unit markings, was "liberated" by aviation machinist's mate, Joseph F. Tucker in the early 1960's. Tucker was a member of Patrol Squadron Two at the time and could not resist a chance to add some of VP 1's property to his wardrobe!

Rear view of Joe Tucker's "VP 1 – #38" G-1 jacket. In addition to its rather practical unit "insignia" stenciled across the back, it shows traces of having once had many name tags and other insignia, now long since removed. Even though Joe was an aviation machinist's mate with VP-2, he could not resist a chance to add a "liberated" garment from VP-1 to his wardrobe. Joe wore this jacket for many years after discharge from the Navy.

A "WAR RECORD" G-1

On first inspection, this G-1 jacket would appear to emulate shades of the "Top Gun" image but it is actually a wearable record of naval service. This is a World War II vintage jacket that has been adorned by a naval aviation veteran with a long and apparently varied career. The jacket is well worn and shows its years of service. Since the jacket has long since parted ways with its owner, its exact history will never be known but its variety of insignia provides quite a base for speculation and theory.

Above: Close up of the right shoulder and chest area of the "War Record" G-1 jacket. There is a small, souvenir type patch for the Naval Air Station at Lakehurst, New Jersey sewn above a Japanese machine made V.P. 42 squadron patch and a U.S.S. Okinawa ship patch on the shoulder.

Right: Close up of the insignia on the left sleeve of the "War Record" G-1 jacket. The goldwire, "bullion" embroidered naval insignia patch appears to have been hand made in a foreign country. The Gemini 4 patch is a commercially sold souvenir type that was probably added years later as a means of recognizing the owners participation with naval forces supporting the operation. The jacket has an Apollo 11 patch, Pensacola patch and Naval Air Station Point Mugu patch sewn to the left chest.

Front view of the well decorated "War Record" Navy G-1 jacket.

Navy & Marine Corps Jackets

RON WILLIS

Ron Willis' G-1 jacket (front and back below) provides an unusual example of a flight jacket worn by non-flying personnel (It should be noted that Lt. Willis only wears the jacket when off duty). Lt. R.L. Willis, USNR (SS) Enlisted, joined the Navy in 1954 and attended Sub School in 1955. He served on the USS Bugara SS331 and the USS Pomfret SS391. Both subs were diesel electrics. Willis was discharged in 1964. In 1971, Ron joined the Naval Reserves as 1st Class Master-At-Arms. In 1982, he was prompted to Chief Master-At-Arms and received his commission in 1985. Since joining the Reserves, Lt. Willis has served with Naval Construction Battalion 22 (Seabees), USS Goldsborough DDG20, Mobile Inshore Undersea Warfare Unit (M.I.U.W.U.) 113, and at the time of this writing is C.O. of Mobile Mine Assembly Group 1711. Lt. Willis' G-1 tells a colorful story of a long career through the insignia of the units in which he has served.

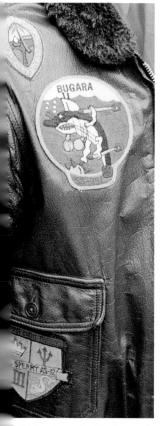

Right front panel with the patch of the USS Bugara and territorial patch of Guam. The lower pocket has the patch of the USS Sperry AS12 affixed, which is a submarine tender.

Left front panel with name plate of Lt. R.L. Willis USNR, and insignia of submarine warfare qualification (enlisted). Although an officer, Willis received his submarine rating as an enlisted man, so his "Dolphins" are silver, as opposed to gold. Also present is the Seabees insignia and the patch of the USS Pomfret.

Subdued insignia of M.I.U.W.U. 113 affixed to left front pocket.

Left shoulder with US flag and Harpoon Missile patches.

USN VR-883

This 1960s vintage nylon flying jacket bears the naval designation of being the jacket part of the winter flying suit. The insulated nylon flying jackets proved to be warmer and much more comfortable than the leather forerunners and these jackets often show a great deal of wear. The insignia is often applied with velcro so that it can be removed for security purposes during flight duty. This is a nice example in an unusually large size.

Above: Close up of the woven nomenclature label in the neck of the nylon winter flying jacket. (Wm. Morris)

Right: Close up of the Asian made USN VR-883 squadron patch sewn to right chest of the nylon winter flying jacket. (Wm. Morris)

Left: Front view of the USN nylon winter flying suit jacket. (Wm. Morris)

"FIGHTING 92 – SILVER KINGS"

This example (below) of the U.S. Navy issue "Jacket, Winter Flying Suit" shows some pretty harsh damage but has survived long enough to make its way into collectors' hands and ultimately some state of preservation. The damage incurred is visually disturbing but not impossible to remedy. The positive aspect is that none of its original insignia has been removed. Once restored this jacket will provide an excellent specimen of a Navy flight jacket from the Vietnam conflict.

Above: Close up of the Japanese machine embroidered VF-92 "Silver Kings" squadron patch sewn to the right chest of the U.S. Navy winter flying suit jacket. Above right: Unusual Asian manufactured VF-92 20th anniversary commemorative patch sewn to the left chest of the U.S. Navy winter flying suit jacket. Note that the patch attempts to recognize the unit's participation and type of aircraft used in both the Korean and Vietnamese conflicts. Right: Detail of the "Sidewinder" missile patch sewn to the right shoulder of the U.S. Navy "Silver Kings" jacket. It is fairly unusual for a flight jacket patch to promote an aircraft's mode of armament but insignia of this type was sometimes issued by the weapons' manufacturer in hopes of promoting acceptance of their product by aircrewmen.

USN-VIETNAM FAIRECONRON-ONE

This example of a Navy issue "Jacket, winter, flying suit" exemplifies anything but winter flying. There must have been a fairly limited use for insulated nylon flying garments in the Southeast Asian climate but several examples of this jacket exist with insignia created for units in the area during the Vietnam conflict. This example reflects a lot of personality and an interesting application of the American flag as a form of insignia.

Left: Right shoulder detail of the Navy winter flying suit jacket. Both the "Viet Nam 1970" tab and "Tonkin Gulf Yacht Club" patch are unofficial insignia typical of the Southeast Asian conflict.

Left: Close up of the insignia of the right chest of the Navy winter flying suit jacket. The "Faireconron One" (Fleet Air Reconnaissance Squadron One) patch is the authorized insignia for the unit but this example was manufactured somewhere in the Southeast Asian area. The signal flag patches represent the owner's initials.

Above: Front view of the Navy winter flying suit jacket. Below: Rear view of the Navy winter flying suit jacket. Note the American flag which is an interesting example of an Asian machine embroidered insignia that is sewn directly to the jacket.

CORSAIR CENTURION

This sage green nylon flight jacket was intended to replace the time honored G-1 leather jacket. Although practical and well wearing, jackets of this type have never eliminated the G-1 from naval inventory. This jacket and its unique insignia reflect the honors and duties of the Navy carrier based fighter pilot.

Front view of the USN CWU type nylon flight jacket with "1000 Hrs" tab and "Corsair II" patches on right shoulder, and the Asian made Atkron (Attack Squadron) 147 patch on right chest. The "Centurion 200-USS Constellation" patch sewn to left chest-note that it appears to have been sewn over the top of the "100" patch. Note also the unusual "Nine Top Hook" patch on left shoulder. (JS Industries)

"BITTER BIRDS"

Although not actually a flight jacket, few things are more emblematic of the fighter pilot's spirit than the souvenir jackets that evolved from the Korean conflict. This example, complete with its Japanese embroidery and reversible feature, is unusual in that its primary design adheres fairly closely to the unit's authorized insignia. So often these unofficial garments are an exercise in individuality and may not even reflect the owner's branch or service, much less his unit.

Left: Front view of the Korean war era USN VF-884 "Bitter Birds" souvenir jacket. Note the USS Boxer Task Force 77 & VF-884 patches on the chest. Also traces of the interior design that can be worn on the back due to the jacket's reversible feature. Above left: Rear view of the "Bitter Birds" souvenir jacket. The entire design has been embroidered directly into the jacket. Above right: Rear view of the "Bitter Birds" souvenir jacket reversed to show the artistic eagle design. The chest of this side has no insignia, only the back does. Perhaps the owner thought this design would be more appropriate for wear "back home." (Wm. Morris)

MARINE AIR WING I – OKINAWA

This novelty jacket is modeled more closely after the WWII armor crewman's jacket but it has been included as an example of unofficial, aviation related jacket. Apparently manufactured somewhere in Asia, this garment reflects strong military styling and versatility. It is, however, doubtful that it was ever worn in a flight duty capacity. Jackets of this type were very popular souvenirs of the Korean and Vietnam conflicts.

Right sleeve of the Marine Air Wing I novelty jacket with fourteen different tabs applied from the shoulder to the elbow!

Close up of the Asian machine made Marine Air Wing I patch sewn to left chest. Unit patches for the Marines ceased to exist officially in 1947 but their designs seem to have lived on and appear in many unofficial capacities.

Close up of the "Short" patch sewn to right shoulder of the Marine Air Wing I novelty jacket. "Short" denoted the owner had little time on the duty station.

Above: Front view of the unofficial Marine Air Wing I jacket. Below: Rear view of the Marine Air Wing novelty jacket. The entire design has been directly embroidered into the material. The lengthy verse, with certain words left out, is fairly unusual and tells quite a story.

CHAPTER VI

Korea to Desert Storm

CLAY E. APPLE

Staff Sgt. Clay E. Apple was a highly decorated veteran of World War II and Korea. Apple piled up an amazing 251 combat missions and was shot down twice – once over the jungles of Burma in WWII and once over the Sea of Japan (Korean War) when he was picked up by the U.S. Navy. His rescue by the Navy inspired him to paint the U.S. Navy NATO flag on the back of his jacket. Apple was credited with seven kills as a gunner, making him an "Ace." The insignia on the left chest of his A-2 is the 98th Bomb Group.

A-2 jacket worn by Staff Sgt. Clay E. Apple in WWII and Korea with 98th Bomb Group insignia painted on left chest of Apple's A-2 jacket. (JS Industries)

Back of Apple's jacket with U.S. Navy NATO flag painted on to commemorate his rescue from the Sea of Japan by the U.S. Navy. (JS Industries)

Unusual American flag painted on right shoulder of Apple's A-2. (JS Industries)

OPPOSITE: Republic test pilot Carl Bellinger climbing into a YP-84 cockpit.

JAMES B. FITZ GERALD

The jackets of James B. Fitz Gerald provide a great legacy of service in World War II and Korea. In World War II, Fitz Gerald flew a B-25 Mitchell named "Stinky" with the 487th Bomb Squadron, 340th Bomb Group, 12th Air Force. During the Korean conflict, Fitz Gerald flew out of Japan with the 8th Bomb Squadron, 452nd Bomb Wing in B-26 Invaders (formerly designated A-26).

Below left: Classic M-41 field jacket worn by James B. Fitz Gerlad while serving with the 487th Bomb Squadron, 340th Bomb Group, 12th Air Force in World War II. The American flag adorns the right shoulder, while the 12th Air Force insignia is on the left shoulder. The shield shaped patch on the right chest is the 340th Bomb Group insignia, and the "Chess Knight" patch on the left chest is the insignia of the 487th Bomb Squadron. Below right: Reverse with painted B-25 "Stinky", and bombs indicating 41 combat missions. (JS Industries)

Above: Front and back of Fitz Gerald's A-2 jacket worn in World War II and Korea. The patches on the front date from World War II, while the painted on "blood chit" on the back is from Korea. Hand tooled leather 340th Bomb Group insignia made on the Isle of Capri sewn to the right chest. Left chest with tooled leather pilot's wing, name tag, and 487th Bomb Squadron patch also made on the Isle of Capri. Early style 12th Air Force insignia in tooled leather on left shoulder. Note the "California" rocker above the insignia and the 340th BG, 487th Sqd. below. (JS Industries)

Korea to Desert Storm

Above: B-15 B flight jacket worn by Fitz Gerald in Korea as a senior pilot. Reverse has a beautifully painted "blood chit." (JS Industries)

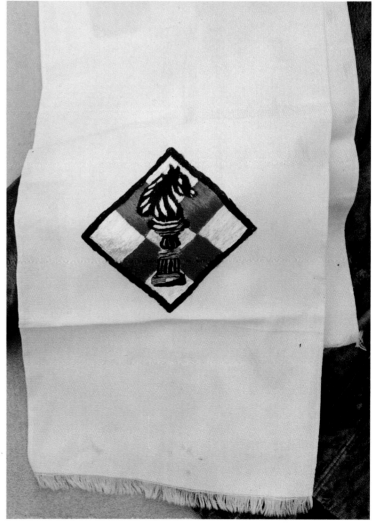

Above: Label of B-15 B and U.S. Air Force winged star insignia. Right: Flying scarf worn by Fitz Gerald with embroidered 487th Bomb Squadron insignia. (JS Industries)

"THE GRIM REAPERS"

The 13th Bomb Squadron, 3rd Bomb Group was widely known in Korea as the "Grim Reapers." They had made quite a name for themselves in World War II with the 5th Air Force and in the Korean War continued their legacy operating B-26 Invaders (formerly A-26), first from Iwakuni, Japan and later from Kunsan, Korea. This unnamed A-2 jacket from the "Grim Reapers" has all insignia painted directly on the leather.

Below left: A-2 jacket worn by a member of the 13th Bomb Squadron, "Grim Reaper," 3rd Bomb Group, in Korea. Below right: Back of "Grim Reaper" A-2. Interesting design of a diving eagle with the tail being that of a B-26 Invader superimposed on a map of Korea. The flash at the top says "13th Bomb Squadron," while the bottom flash is no longer legible. (JS Industries)

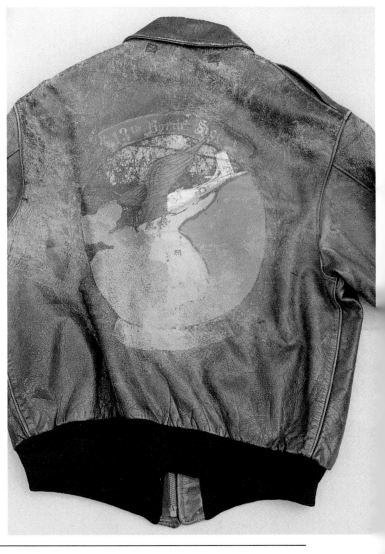

452nd BOMB WING

This beautifully painted A-2 jacket on display at the U.S. Air Force Museum was worn by an American airman who flew 55 missions on a B-26 Invader with the 452nd Bomb Wing in Korea. There are 30 bombs in the top group under the B-26 and the lower bombs are divided into groups of five, each over the following targets; Pyng Yang, Sinanju, Sunchon, Yangdok, and Hamhung. (AFM)

Korea to Desert Storm

Below: Front and back of B-15 D jacket worn by Sam Nunez, 75th Air Depot Wing, 5th Air Force. (JS Industries)

SAM NUNEZ

It is very unusual to find a B-15 D with painting, but Sam Nunez's jacket provides a spectacular example from the Korean war era. Nunez served with the 75th Air Depot Wing, "Fighting 75", 5th Air Force.

KOREAN WAR FLIGHT VEST

This vest is constructed of olive green cotton with a brown alpaca lining and zippered closure front. No labels remain but it appears to be a World War II vintage Army Air Force issue vest. It was not uncommon for World War II surplus clothing to be reissued in Korea. The presence of recognition oriented insignia of this vest would seem to indicate it was possibly used as an outer garment by an air crewman during the Korean conflict.

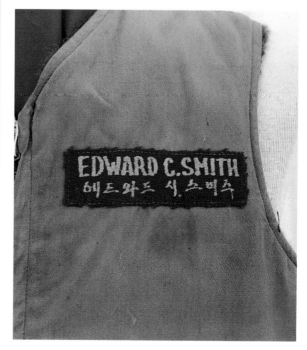

Front and rear views of the Korean War era flight vest. Note tattered remains of allied flags patch sewn to the back of the Korean War era flight vest. The upper flags are obviously the U.S. & U.N. designs. The lower flags are the flag of Great Britain and what appears to have been an inked or painted rendition of the South Korean flag now all but completely faded away. Right: Detail of the embroidered name tag sewn to left chest of the Korean War era flight vest. The embroidery style and type of coarse woven wool are typical of Korean manufacture by native tailors. The Korean characters denoting owner's name are also obvious clues of the insignia's origin.

37TH BOMB SQUADRON-KOREA SOUVENIR JACKET

Like other jackets shown in this work, the 37th Bomb Squadron souvenir jacket is so typical of the period it evolved from we felt it must be included to perpetuate the spirit of the men who formed the unit it represents and to exemplify what could be done outside the area of regulated jurisdiction. This example is extremely colorful and bears some beautiful oriental embroidery work. It is completely reversible but is unusual in the fact that the reverse side presents no change in color, which is usually the point of making a garment reversible.

Pilot, Major Robert Price (second from right) and crewman of the 37th Bomb Squadron, 17th Bomb Wing, receiving a commendation for operational flight safety on night combat missions during the Korean war. (Col. Robert Price U.S.A.F. ret.)

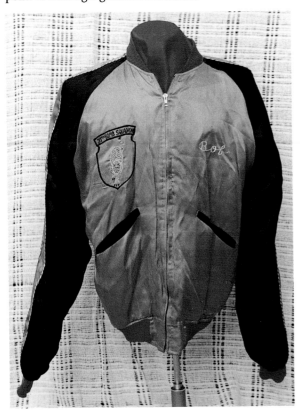

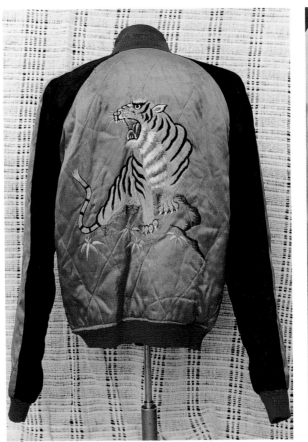

Above left: Rear view of the Korean War 37th Bomb Squadron souvenir jacket reversed to show ornate, silk embroidered tiger design on back. Above right: Detail of the "Bengal Tiger" Squadron insignia directly embroidered into the right chest of the Korean War 37th Bomb Squadron souvenir jacket. Note the striking detail and color that are so typical of oriental work of the period.

Left above: Front view of the Korean War USAF 37th Bomb Squadron souvenir jacket. Left: Rear view of the Korean War 37th Bomb Squadron souvenir jacket. Note the 3-flag design with the Japanese flag in center. This probably indicates the country of origin. The unit's unique aircraft, complete with recognizable markings is interesting.

Korea to Desert Storm

UNUSUAL LEATHER USAF FLIGHT JACKET

This unique specimen is actually a well-worn, WWII vintage USN issue G-1 leather flight jacket, but at some time after the war it changed branches of service! The jacket is well worn and shows repairs such as elbow patches and large reinforcing panel across shoulder area of the upper back. The USAF patches now sewn to it appear to be 1950's era due to their size and construction. With no background information, the possibility does exist that this jacket was created for wear by an admiring young aircraft fan.

Right: Front view of a well-worn WWII issue USN G-1 leather flight jacket converted for use by a USAF member some time in the 1950's. Above left: Close up of the large, machine embroidered on cotton twill "U.S. Air Forces in Europe" patch on the right chest of the WWII issue USN G-1 jacket. Above right: Close up of the shoulder/pocket size USAF "50th Tac Ftr. Wg." patch on the left shoulder of the WWII issue USN G-1 leather jacket.

Left: Front view of Major General Stanley C. Beck's blue cotton flight jacket. Note Command Pilot Wing, Basic Parachutist badge and Missile Badge direct embroidered into General S.C. Beck's cotton flight jacket. This style and type of embroidery work is typical of native craftsmanship in a number of oriental countries.

MAJOR GENERAL STANLEY C. BECK

This blue cotton flight jacket is patterned directly after the nylon constructed L-2B jacket but it is not a standard issue item. This particular example was issued to then Brigadier General Stanley C. Beck and is a uniquely personal specimen. General Beck had a distinguished career in Southeast Asia and went on to attain the rank of Major General. His qualifications in the "hands on" work of the Air Force are reflected by the wings and missile badge direct embroidered into the left chest of his jacket.

Right: Typewritten data label inside Major General (then Brigadier) Stanley C. Beck's cotton flight jacket. Labels like this provide fairly solid evidence of ownership and often help date the garment. Due to washing and oversight, they are often unreadable, incomplete or missing!

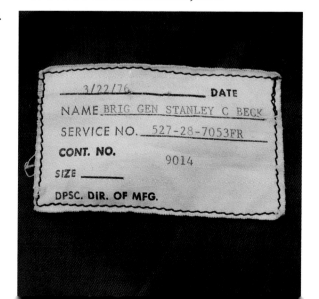

MAJOR BURKE

The MA-1 jacket has gone through similar variations in production as the L-2B. The MA-1 was a heavier insulated example for colder flight conditions and lacked some of the details of construction of its lighter counterpart. This is a relatively late example as very few changes have taken place in its design since the addition of the bright orange lining. The size, placement, and type of insignia dates this jacket in the 1960-70's era.

Front view of Major Burke's MA-1 flight jacket. Note the bright "indigo" orange lining. (JS Industries)

Detail of insignia on the left shoulder, chest and sleeve of Major Burke's jacket. The "SAC Mach 2+ FB-111" patch very definitely dates the period of use for this specimen. (JS Industries)

Detail of chessboard unit insignia on the right shoulder of Major Burke's jacket. (JS Industries)

MA-1 B-1B TEST CREW

This late issue USAF MA-1 flight jacket is an interesting example of flight gear used by a crewman involved in testing of a proposed operational aircraft – the B-1B bomber. Perhaps the most controversial aircraft of all time, the B-1B fell victim to friendly fire of a sort and became a legend in its own time. The identity of the original owner has been lost.

Front view of the B-1B test crewman's MA-1 jacket. The B-1B "Starburst" patch on the left chest (below left) was probably distributed to the crews by Boeing. Below right: Detail of small American flag and Ellsworth Air Force Base B-1B patch sewn to right chest of the crewman's MA-1 jacket. The presence of the patch suggests the jacket may have been worn by a civilian technical representative for Honeywell.

LT. GENERAL GABRIEL, U.S.A.F.

This standard issue, L-2B "Light Zone" flight jacket has only one distinction – it was issued to a general officer and still bears the insignia to prove it! This jacket is a very late example in the dark shade sage green and all its unit related insignia is affixed with velcro so that it can be removed easily. Lt. General Gabriel had a record of distinction with the Air Force during the Vietnam conflict and some of the insignia on this jacket appears to have had its origin in this part of his career.

Above: Front view of Lt. General Gabriel's L-2B "Light Zone" flight jacket. Below: Detail of insignia on the left shoulder, sleeve and chest of Lt. General Gabriel's L-2B jacket. Note the mixture of both color and subdued insignia. The cloth name tag with command pilot wing appears to have been made in the Southeast Asia area. (JS Industries)

Korea to Desert Storm

RALPH M. JEROME

This dark blue nylon L-2A flight jacket represents the earliest variant of this intermediate weight nylon flight jacket. Introduced in the early 50's, the L-2A was accompanied by a wide range of other nylon flight garments, all produced in the dark blue shade which was apparently short lived compared to its predecessor – sage green. The L-2A was a direct descendent of the B-15 jacket of World War II. The basic construction changed little, but the fur collar, reminiscent of the early days of flight, was gone for good.

Left: Front view of the L-2A worn by Colonel Jerome. Note the woven specification label in the neck. Of interest are the Colonel's rank devices (on shoulders) and the leather name tag on the left chest, which are silver leaf printed on black leather. The large, unmarked 48th Fighter Group patch on the right chest is typical of 1950s squadron insignia due to its size and lack of unit designation. The smaller Air Defense Command patch on the left shoulder appears to be of later vintage and may indicate Colonel Jerome enjoyed the use of this jacket well after its period of issue. (JS Industries)

The L-2B intermediate flight jacket was produced in three basic variations from the 1950's to recent years. The latest variant is represented by this example worn by a U.S.A.F. Colonel in the 48th Fighter Group sometime in the 1960's. This L-2B is the standard, lightly insulated construction and is reversible to bright orange for recognition from the air in the event of forced landing. The bright orange is officially "indigo" orange but it has been referred to as "international" orange.

Left: Front view of Colonel Ralph M. Jerome's L-2B flight jacket. The bright orange lining is clearly visible in this photo. The silver leaf impressed name tag actually slips into a clear plastic pocket on the left chest and is easily removable for security purposes or in the event of reissue. Plastic sealed colonel's eagles have been stitched to the shoulders and embroidered patches for the 48th Fighter Group and Air Defence Command can be seen on the right chest and left shoulder. (JS Industries)

ARMY NOMEX FLIGHT JACKET

The need for flame retardant flight clothing was fulfilled with the introduction of Nomex, and this dramatic new material found its way into the construction of many new garments for flying crewman of all service branches. This Army flying jacket proved to be not only a fire-safe piece of clothing, but a comfortable and popular garment among its recipients. Lightly insulated, it blended well in design with the tendencies of Nomex to be a slightly warmer material than cotton. The jacket was never overly popular in Vietnam, but its proficiency as an intermediate outer garment would be proven in the United States and other foreign areas of operation. This example has seen enough use that its original darker shade of olive green has begun to fade. This was typical of the Nomex material when exposed to laundering and long periods of use in sunlight.

Right: Front view of an Army Nomex flying jacket with permissible, subdued cloth insignia.

Below: Detail of the left chest and shoulder area of the army nomex flying jacket. Note the Sr. Aviator and Airborne Qualified badges on the chest, captain's rank bar on the shoulder and the 101st Airborne Division patch on the shoulder. The upper half of the zippered sleeve pocket, with provision for four pencils, can also be seen in this photo.

Below: Detail of the U.S. Army/NATO size and specification label sewn inside neck of the Army Nomex flight jacket.

"THE DEANS" NOMEX FLYING SHIRT

The aviators age-old fears of injury or death inflicted by aviation fuel or the burning aircraft itself were finally addressed in the form of a flame retardant material during the Vietnam war. The introduction of flame retardant Nomex material brought with it changes in types and styles of flying garments. One of the most interesting was the Army's introduction of the two piece flight suit. The nomex shirt was accepted among Army aircrews but was often criticized because it was not well suited for the heat of Vietnam. The baggy, unpressable, pocket besieged trousers were often criticized for their "unmilitary" appearance but the Nomex flight suit saw Army air crews through until the end of the conflict in Southeast Asia. Shirts, such as this example, became "bulletin boards" for the wide variety of insignia that became available to G.I.s through the efforts of enterprising native craftsman.

Left: Close up of the left chest and shoulder area of the Army Nomex flying shirt. This subdued insignia was produced in mass by native tailors and examples flourished during the later years of the Vietnam conflict. Visible in the photo are the warrant officer's branch device on the collar, Army Aviator's wing and "U.S. Army" tab above the pocket seam and a 1st Aviation Brigade patch on the left shoulder.

Front view of an Army issue nomex flying shirt. This shirt was the upper half of a two piece flight suit introduced by the Army with the acceptance of the flame retardant Nomex material.

Detail of the 120th Aviation Company patch sewn to right chest pocket of the Army Nomex flying shirt. Insignia embroidered in only shades of black and green were known as "subdued." Usually any subdued patch had a colorful counterpart but the subdued versions were obviously more practical on a combat garment due to the lower level of visibility.

CLAUDE A. "CHUCK" BERRY

This standard issue jungle fatigue jacket was issued to then Major Claude A. Berry while flying helicopters in Vietnam. Berry was mentioned earlier in this book and has the distinction of being a 3-war Army aviator: glider pilot in World War II, a liaison pilot in the Korean War, and a helicopter pilot in the Vietnam War.

Right: Front view of Major Claude A. Berry's Army issue jungle fatigue jacket.

Left below: Detail of the 120th Aviation Company patch sewn to right chest pocket of Claude Berry's jungle jacket. This patch was hand embroidered by native sources in Vietnam and is a typical example of the unlimited unofficial insignia that was so prolific among aviation units in the war.

Right below: Close up of the left chest and shoulder area of Claude A. Berry's jungle jacket. The Army aviator's wing insignia has been hand embroidered by a Vietnamese native. The "Razorbacks" aviation company patch and 1st aviation brigade insignia (visible on pocket and shoulder respectively) are also examples of native craftsmanship. This type of insignia was very popular with helicopter crewman and was accepted by commanders as a source of morale and pride in unit.

Left: Front view of an Air Force issued Vietnam jungle fatigue jacket used by a navigator in the 366th Tactical Fighter wing. Of note is the "In-Country" machine made 366 TFW patch sewn to the left chest pocket. The cloth Lieutenant bars, wing and name tapes were also manufactured somewhere in Vietnam. Right: Detail of the locally made US Air Force 366 Tactical Fighter wing patch sewn to left chest pocket of the Air Force jungle jacket.

VIETNAM JUNGLE JACKET

The jungle fatigue jacket was the standard issue outer garment for every service branch of the United States forces operating in Vietnam. This example was issued to a navigator assigned to the 366th Tactical Fighter wing stationed at Danang. The mixture of subdued insignia with a color unit patch is not unusual but the unaltered sleeves are an exception for an Air Force issue example. It was very common for Air Force personnel to have the sleeves cut short for coolness and comfort.

MELVIN A. McDUFF

Colonel Melvin A. McDuff was called to active duty in 1940 and came home from WWII a full colonel. After returning to civilian life for a few years he was re-active for the Korean War. Colonel McDuff functioned as the Deputy Post Commander at the Army Aviation School, Fort Rucker, Alabama from 1962 to 1964. While there he was issued an MA-1 insulated flight jacket. Colonel McDuff was never an Army aviator so the insignia is relatively basic, as might be found on a fatigue shirt.

Right: Front view of the MA-1 flight jacket used by Colonel Melvin A. McDuff – Army of the United States. Left: Detail of left shoulder area of Colonel Melvin A. McDuff's MA-1 flight jacket. Note the embroidered colonel's eagle sewn in line with the shoulder seam and the Army Aviation School patch with Aviation Center insignia below. The zippered sleeve pocket with pencil loops can also be seen in the lower center of the photo.

"APACHE MIKE"

This private purchase A-2 leather flight jacket is an unusual addition to our book and a very timely one. Although the U.S. Air Force has officially reinstated the leather A-2 leather flight jacket, the army, as of yet, has not. This jacket appears to have been decorated, worn, and brought home by a Desert Storm veteran who flew AH-64A "Apache" helicopters for the Army. It is perhaps a striking reminder that nothing has changed the spirit of the men who fly combat aircraft, regardless of the type of craft or the era from which it is born.

Detail of the left chest of "Apache Mike's" private purchase A-2 jacket. Note the silver leaf impressed leather name tag and the issue style 3rd corps patch. Stitch holes in the area of the name tag indicate the name tag and patch now on the right chest have been moved to this present location for some reason.

Left: Front view of the Desert Storm vintage Apache helicopter pilot's unofficial A-2 jacket. Below left: Rear view of the Desert Storm A-2 jacket with its hand painted war art design. It is strikingly reminiscent of those who followed before it 50 years past. The "Ain't Outta Luck" may be a personally preferred phrase of the jacket's owner, or something that relates to a particular time of incident.

Right: Detail of the right shoulder and chest of "Apache Mike's" Desert Storm vintage A-2 jacket. The shoulder insignia has been neatly handpainted directly on the leather. The AH-64A patch is a machine embroidered example that has been sewn in place. "Generic" aircraft insignia are often distributed among units by the manufacturers and this may be an example of that. Unofficial insignia pertaining to aircraft and flying units are not unusual, especially in times of activation for combat operations.

"HOG HEAVEN"

This jacket, scarcely two years old at this writing, exemplifies the airmen's love of his aircraft and the desire to be unique in an otherwise "uniform" environment. Little is known of the individual who created this interesting specimen but it is perfectly viable to theorize that he was an admirer of those many airman who preceded him in past wars and their individual attempts at self expression on clothing and aircraft.

Close up the 354th Tactical Fighter Wing patch sewn to right shoulder of the USAF CWU type jacket. Note the A-10 aircraft and "Desert Shield" motif.

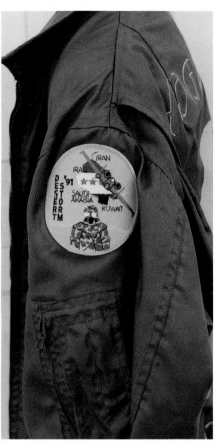

Close up on the unofficial A-10 aircraft "Desert Storm" patch sewn to left shoulder of USAF CWU type jacket.

Front view of current issue USAF CWU type name tag on the velcro panel on left chest.

Close up the 76th Tactical Fighter Squadron patch sewn to left chest of the USAF CWU type jacket.

Rear view of the USAF CWU type nylon flying jacket with colorful art work reflecting the rather glamorless nickname of a fantastic combat aircraft and its role in the recent Middle East conflict "Desert Storm."

AMERICAN FLIGHT JACKETS, AIRMEN & AIRCRAFT

These jackets are truly a tribute to all the airmen who went off to war. They were young, products of the Depression, but instilled with an American spirit of "can do." Many of them, who had been riding around on bicycles the year before the war, were now flying and manning highly advanced aircraft. They epitomized the spirit of American ingenuity, improvisation, and organization, turning a tiny out-dated Air Force in 1941 into the largest one ever to be known; one that would come to dominate the skies over Europe as well as those of the Far East.

In virtually all cases I found these men to be the most effacing and gracious of human beings. Some did not even want their names mentioned, since they felt they did a job that everyone else had performed.

Joe Jacoby was a staff Sergeant with the 90th Bomb Group, 319th Squadron, 5th Air Force. He was a ball turret/radio operator and sometimes even had to act as toggle. According to Jacoby, all crewmen on a B-24 had to have some knowledge of other jobs on the plane. The insignia, Asterperious on the left chest and the 90th Jolly Roger on the right, are Australian made. Joe Jacoby related that the Jolly Roger B-24 painted on the back of the jacket, entitled "Mary," was painted on the Island of Ieshima. In the Pacific theatre, it was all too common to see paintings of women or their names on the sides of bombers. Air units in this theatre were truly cut off from the outside world, and young air personnel had but one thing on their minds during idle hours...

Most had not really thought back to or talked about their "stories" since the war, and in most cases it was cathartic and emotional. Tom Brokaw was not mistaken in labeling the men and women of those times as "The Greatest Generation."

I would like to thank several individuals who were very helpful to me in my quest. Robert H. Powell Jr. of the 352nd F.G. ASSN, George H. Menzel of the 401st Bomber Group Association, Lyman Goff, of the 491st Bomber Group, Robert F. Wittling of the 449th Bomb Squadron, a native of South Bend Indiana, who knew my father, Edward Longhi, a 1938 All-America Center at Notre Dame, who played for Elmer Layden. All of them and many others, true sons of America.

An early shot of "The Dude" in the South Pacific before the famous 90th BG Group insignia was applied.

New Photo Gallery (Courtesy of Leighton Longhi)

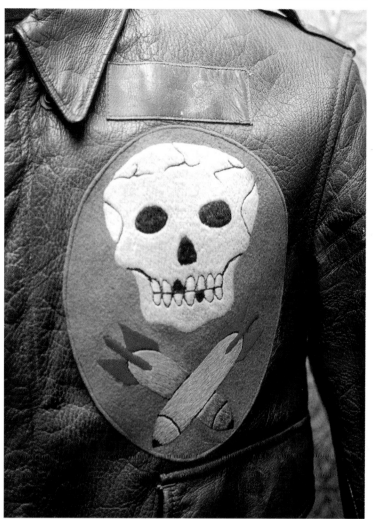

A shot of Boiteau wearing his jacket with Gentry's crew.

Sergeant J.A. Boiteau was a 90th Bomb Group photographer who served aboard Lt. William Gentry's B-24, "The Dude." Boiteau's jacket features a fabulous and possibly unique oversize Australian made 90th Bomb Group patch.

Photo by Boiteau, for which he received the Air Medal on his First Mission when their B-24, piloted by Gentry, sunk a Japanese freighter in the Bismarck Sea. (If you look closely you can see Japanese crewmen crouching on the front of the ship.)

A shot by Boiteau of Bob Hope wearing a 90th BG A-2.

Left: A color shot of Boiteau after receiving the DFC.

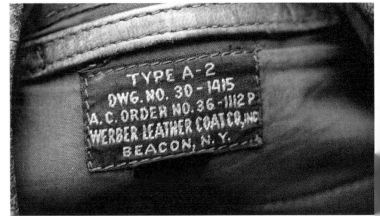

J.B. Bass was an airman with the 1st Combat Cargo, 1st Squadron. The A-2 is made by the Werber Leather Coat Co. of Beacon, NY, a very rare maker, with an extremely early date of 1936. Unfortunately, nothing else has been researched about Bass, who was most likely an early member of the Army Air Corp.

A picture of a multipiece leather American flag with leather blood chit below. Both shoulder insignia—China Burma India and Air Force insignia—are theatre made.

A beautiful view of Bass' A-2 with theatre-made multipiece leather insignia. Typical of early jackets, reinforcing studs can be seen at the bottom of the jacket near the end of the waistband.

New Photo Gallery (Courtesy of Leighton Longhi)

Lt. Douglas D. Stewart was a P-38 pilot with the 343rd Fighter Group, 54th Squadron, Eleventh Air Force, stationed in Attu and later Shemya in the Aleutians. Stewart also flew F-86s in Korea and later served in Vietnam, retiring from the Air Force as a Lt. Colonel. The 343rd Fighter

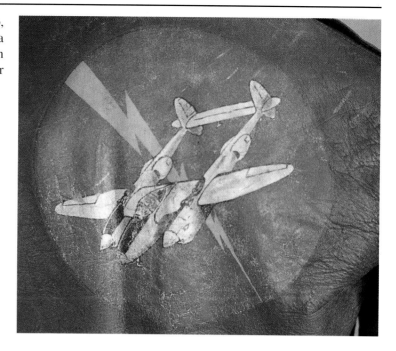

Left: Front view of Stewart's A-2 with a chenille patch representing the P-38 school at Santa Maria Air Base in California. One can see a yellow jacket with a "lightning bolt" studying a book.

Below, Left: A period photo of Stewart in his P-38. (via Stewart)

Back of Stewart's A-2 with Roundel painting of a P-38 straddling a lightning bolt, the sobriquet for this aircraft. Stewart remembered that a fellow pilot in the 54th painted his jacket.

Group was originally commanded by John S. Chenault (Clare Chenault's son) and converted from P-40s to P-38s. Another famous son, William Mitchell, Jr., flew with the 394th Squadron of the same group. During his service Stewart received the *Legion of Merit,* DFC, and Air Medal with oak leaf clusters.

A period photo of Mitchell in his P-40.

Boise Bee

Lt. Duane Beeson was the leading ace of the ETO with 17.3 kills when he was shot down on April 5, 1944, strafing an airfield outside of Berlin. He was also the ranking P-47 ace of the famous 4th Fighter Group and served with the 334th Fighter Squadron. Beeson, like many 4th Fighter Group pilots, had previously served with the RAF, joining the Royal Canadian Air Force in 1941. In England he was assigned to the Eagle Squadron (no. 71) and was transferred to the 4th Fighter Group in October of 1942. He achieved his first victory on May 18, 1943, when he shot down an Me 109 near Ostend, Holland.

Right: A detail of Beeson's left sleeve with an early style 8th Air Force painted insignia.

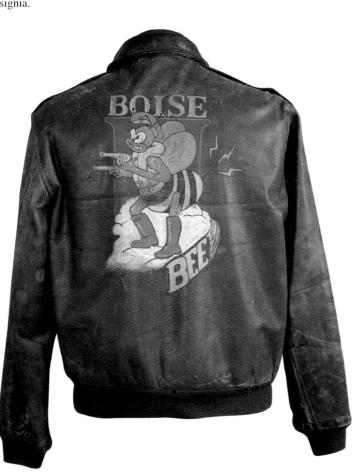

Beeson, a native of Boise, Idaho, had all his mounts (P-47s as well as P-51 painted with the design of a gun-toting "Boise Bee" cowboy.

New Photo Gallery (Courtesy of Leighton Longhi)

Lt. (later Major General) Cuthbert A. (Bill) Patillo

Bill Patillo flew 135 combat missions for the 352nd, known as "Bluenosed Bastards of Bodney." It was said that the 352nd got its name from Hermann Goehring, who said the war was lost when the "Bluenosed Bastards of Bodney" appeared over Berlin. Patillo was credited with shooting down an Me 262 on April 16, 1945, and became an ace in a day (ground) when he destroyed six German aircraft. Unfortunately for Patillo, on his last strafing run over the airfield he was shot down by enemy anti-aircraft fire. During WWII he received the DFC and Air Medal with two oak leaf clusters. Patillo went on to have an illustrious career with the Air Force, retiring as a Major General. His commands and decorations are too numerous to mention in this writing. The two Patillo brothers were only one set of two twins to serve in the Eighth Air Force as fighter pilots.

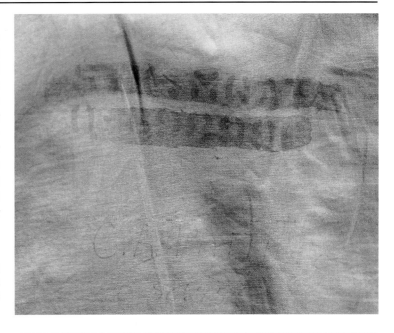

Right: Interior of Patillo's jacket shows both Patillo's name and that of Sanford Moat, another ace of the 352nd with 8.5 aerial victories. Jackets were always recycled until the end of the war, and Patillo received Moats' jacket when he joined the 352nd. Moats also went on to become a General in the Air Force.

Close up view of one of the most famous of the Fighter Squadron patches, a baby with a flight helmet holding a machine gun, and a riding crop. The riding crop was the result of a stateside order by its Commander, J.C. Meyer, that pilots carry a crop to boost morale. Sam Perry designed the insignia, and pilot Lt. Karl Waldrow posed in diapers as "the little bastard."

A view of Lieutenant Patillo's A-2 with the Squadron insignia of the 487th.

Lt. Patillo in front of his P-51 "Sweet and Lovely."

A picture in 1953 of Bill Patillo (left rear) with his twin brother, Buck (right rear), founders of the famous Air Force demonstration team "The Thunderbirds."

Interior view of Mathies' A-2 with his stenciled name.

Archibald Mathies - Medal of Honor Winner

Archibald Mathies and his family emigrated from Scotland to the United States, settling in Pennsylvania. At the outset of the war Mathies enlisted in the Army Air Force as a mechanic and was later trained as an aerial gunner. In December of 1943 he was sent to England and assigned to the 510th Bomb Squadron of the 351st Bomber Group, 8th Air Force. He served on a B-17 named "Ten Horsepower" (a play on a B-17's ten crewman). While flying his second mission on February 20, 1944, to Leipzig his aircraft was seriously damaged by fighter attacks. The attack killed the co-pilot and rendered the pilot unconcious. Mathies, without virtually any flight experience, entered the shattered cockpit and managed to fly the aircraft back to its base, where he had the crew bail out. However, he refused to leave the unconcious pilot and attempted to land the aircraft with a B-17 sent up from the base, the only chance to save the pilot. Unfortunately, on his third attempt the bomber stalled and crashed. February 20th was to be the only day in the history of the Eighth when more than one medal of honor was awarded. One to William Lawley, and the other to the two crewmen of "Ten Horsepower," Archibald Mathies and Walter Truemper.

Close-up of Mathies' name above the 351st patch.

An overall view of Mathies' jacket with his name tag above the group's insignia of the 351st.

New Photo Gallery (Courtesy of Leighton Longhi)

1st Lieutenant William H. Allen was assigned to the 343rd fighter squadron of the 55th Fighter Group, Eighth Air Force. He flew both P-38s and P-51s. During Allen's WWII service he received the Air Medal with 4 Oak Leaf clusters and the DFC with 2 Oak Leaf clusters. On September 5, 1944, while involved with a flight of the 343rd, Allen ran into a flight of 16 German aircraft over Soppingen airfield. Allen was shot down after claiming 5 German aircraft, making him an ace in one day.

Right: Photo of Allen's B-15A jacket.

Photo of Allen in front of his P-51, "Pretty Patty II."

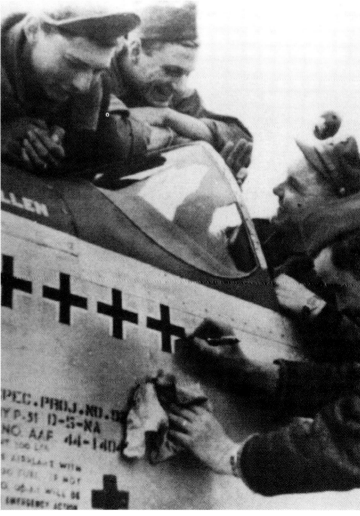

Photo with Allen and his crew while his tally is being painted on his aircraft.

"Dolly's Sister"

Sergeant Gino Bartolino was a radio air gunner with the 453rd Bomb Group, 734th Squadron of the 8th Air Force. He flew 30 missions with the 453rd from 1944 through 1945. His A-2 jacket sports the name "Dolly's Sister" and has a painting of a 453rd B-24 with 15 bombs painted on the back (each bomb counting for two missions). An interesting footnote to the 453rd was that its operations officer was the academy award winning actor Major James Stewart (later Lieutenant Colonel).

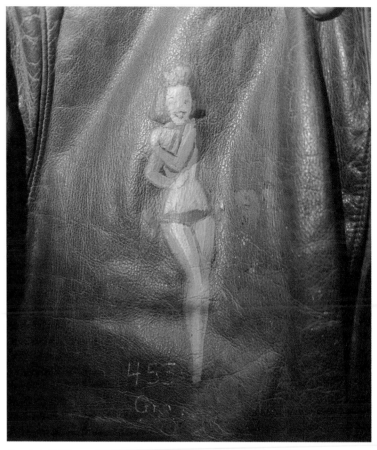

Photo with Allen and his crew while his tally is being painted on his aircraft.

The front of the jacket "bares" a bikini clad girl with "453rd Bomb Group" below. On the right chest is Bartolino's name tag, with a blue square signifying 4 combat airman.

Detail of "Dolly's Sister" with a 453rd B-24.

New Photo Gallery (Courtesy of Leighton Longhi)

1st Lieutenant Richard Paul Casterline was a B-24 bombardier with the 448th Bomb Group, 713th Squadron, based at Seething, England. He had attended art school in Cleveland before the war and was considered by his fellow airmen as the "old geezer," being 26 years of age at the time. He flew 32 missions, his first on December 30, 1943, to Ludwigshaven and his last one on June 18th to Hamburg. He also took part in five D-Day related bombingss. All the art work on Casterline's jacket was done by him, and he related that all the squadron and group insignia were designed by Bob Harper in Special Services.

Photo of target Brunswick, Germany, April 8, 1944. Casterline related that Brunswick was one of their toughest targets, with heavy enemy fighter and flak defense. Their final run was marked by "panic and confusion," with planes at all altitudes dropping bombs which nearly hit other (including his) planes in the bomber stream. On that day, the defense was so intense that Jimmy Stewart, who led the overall formation, had to bravely bring it back three times in an attempt to get it right.

According to Casterline, Harper designed this winged eagle-like animal diving out of the U.S. insignia with red lightning bolts as the group's insignia. The name of their aircraft "Cold Turkey" (they first landed in England on Thanksgiving) was not added to the jacket.

Picture of Colonel Judy decorating Lt. Casterline with the Distinguished Flying Cross, May 1944, Seething, England). On Casterline's 15th mission his aircraft was badly shot up, with the pilot and navigator wounded. The co-pilot managed to nurse the aircarft back to a crash landing at Seething. Casterline was one of the few without wounds, and he flew on other aircraft until his crew returned (the reason for extra missions).

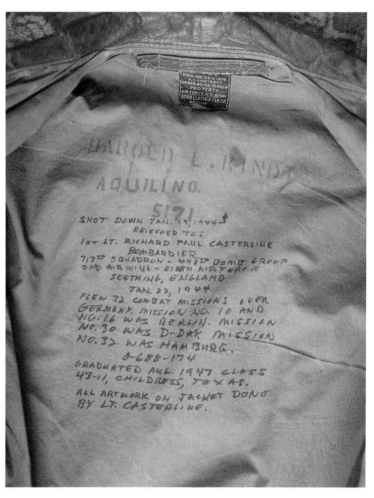

Paul Casterline inherited his jacket, and according to him, wore it on all of his 32 missions.

AMERICAN FLIGHT JACKETS, AIRMEN & AIRCRAFT

Paul Casterline inherited his jacket, and according to him, wore it on all of his 32 missions.

Paul Casterline inherited his jacket, and according to him, wore it on all of his 32 missions.

This beautifully designed jacket has 32 bombs with Berlin and D-Day missions highlighted. On the right chest is the insignia of the 713th Squadron, a musketeer-like figure descending on a bomb with a machinegun.

Seargeant Jack C. Hinkle was an armourer/gunner assigned to a number of 91st Bomb Group squadrons. The 91st flew B-17s out of Bassingbourn, England.

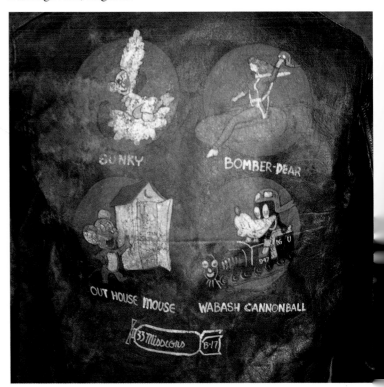

Hinkle served on four famous 91st Bomb Group aircraft: "Bunky" (323rd) "Bomber Dear" (322nd); "Out House Mouse" (323rd); and "Wabash Cannonball" (322nd). This jacket is perhaps one of a kind, as there are no other known jackets portraying several aircraft.

New Photo Gallery (Courtesy of Leighton Longhi)

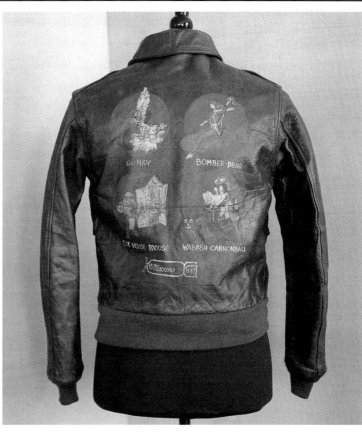

On the front of Hinkle's jacket is the Squadron emblem of the 324th (where he started out), painted on leather. George Odenwaller, another crewman from "Out House Mouse," recounted that Tony Starcer did virtually all the nose art and jacket paintings in the 91st. Starcer is certainly the most well known of the Eigth Air Force artists, and his creations, such as "Out House" and the "Memphis Belle," will always be the most famous. Another jacket, also owned by Hinkle, "Sleepy Time Gal," has the insignia of the 323rd on the front.

Jack Hinkle (first person standing on right) in front of a 91st Bomb Group B-17.

"Ten Fakers"

Eighth A.F. 351st Bomber Group 510 B.S.—The 351st flew B-17s out of Polebrook, England. Sergeant Percy Ruis, a native American, was a 2nd Engineer and Ball turret operator on "Ten Fakers." He received the DFC and Air Medal and was credited with four confirmed German aircraft shot down.

The "Ten Fakers" is painted in gothic style—very common to this group—with a black "J" on a white triangle (dating the insignia prior to March 1944), when the unit used olive-drab aircraft. 25 Bombs signifies 25 missions, which was the limit earlier in the air campaign against Germany. "Ten Fakers" as an aircraft name was a none too subtle connotation with which all ten crewmen tried not to acknowledge the sheer terror of air warfare over the European Continent.

Left: On the left chest pocket is a wool patch of the 351st Bomber Group, and on the right is the Squadron insignia of the 510 B.S.

Crew photo of the "Ten Fakers," with Percy Ruis in the back row, third from the left. As can be seen, Ruis was the perfect size for the terribly restrictive and claustrophobic space of the ball turret.

New Photo Gallery (Courtesy of Leighton Longhi)

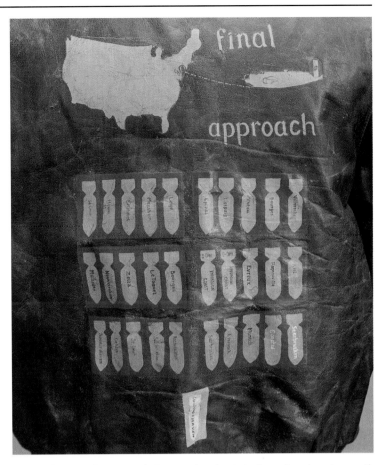

Eagen's A-2 was beautifully designed with a strong color sense, highlighting 31 bombs and targets. It is somewhat rare and certainly historically interesting, with the last mission to Ludwigshafen painted on a yellow ground. At the top is a 458th B-24 flying towards NYC (New York City). "Final Approach" was perhaps a sentiment shared by all aircrewmen the minute they set foot in their theatre of operations.

"Final Approach"

458 Bomber Group, 752nd Bomb Squadron, 8 AF ETO. Based at Horsham St. Faith. James Eagen was a Staff Sergeant who flew on a B-24 named "Final Approach." He was a tail gunner and knew that he could make his final approach when his tour ended in August of 1944. A photo of "Final Approach" and its crew can be found on p. 161.

Sergeant Earl Sivey was an armourer/gunner flying on B-24s with the 445th Bomber Group, 702nd Bomber Squadron based at Tibenham, England.

In two of the photos Sivey is seen with his painted A-2 jacket before the bombing missions were added. Such photos, actually showing the airman in theatre with the jacket already painted, are prized by collectors.

Close-ups of Sivey's jacket with a very patriotic painting of the battle flag of the Republic surmounted by a winged eagle. This iconography originally dates back to the time of the American Civil War. 32 bombs are now painted on the back of Sivey's jacket, signifying the end of his duty. Often the artwork was done first, and all the bombs added at the completion of duty!

"Our Baby"

Sergeant Bertram Chernow was an armourer/waist gunner on a B-24 entitled "Our Baby," which was affectionately named by their crew. Chernow was in the 448th Bomb Group, 715th Squadron, based at Seething, England, and served on "Our Baby" from June 13, 1944, till March 3, 1945.

Chernow picture of "Our Baby," with Chernow kneeling, first person on the left.

A picture of Chernow in dress uniform.

New Photo Gallery (Courtesy of Leighton Longhi)

A fabulously painted A-2, "Our Baby" is done in large gothic letters above a B24 of the 448th. The lower portion of the jacket with 30 bombs denotes thirty missions, the later requirement for heavy bomber crewmen. In between the bombs, another "baby" is posturing in diapers.

"Time's Awasting"

1st Lieutenant Arthur Case was a Bombardier on a B-24 entitled "Times' Awasting," assigned to the 855th Squadron of the 491st Bomb Group based at North Pickenham, England.

Front of jacket shows the 855th Squadron painted on the left breast pocket of the jacket. This seemed to be common in the 491st, where Squadron insignia appear where the group insignia usually is placed. Perhaps, because no known patch of the group has ever been discovered.

This beautiful jacket, in almost unused condition, features a pin-up girl starting to unfasten her under garments. Below is a message that this girl believes "Time's Awasting." The upper portion of the jacket has an ammunition belt-like painting of 30 bombs with a super imposed "European theater of Operations."

AMERICAN FLIGHT JACKETS, AIRMEN & AIRCRAFT

Picture of the crew of "Time's Awasting." (Case seated in the middle of the first row).

"Unconditional Surrender." This is a fascinating footnote on how the same images were used and copied with different titles. Lt. Col. Lyman Goff was the Deputy Group Commander of the 491st. Goff was involved in the initial formation of the group and oversaw the group's flight from the U.S. to North Pickenham. "Unconditional Surrender" was Col. Goff's aircraft. Goff's nose art must have been popular with the group, as "Time's Awasting" suggests. (Goff can be seen standing, second from the right).

"Shack Rabbit"

"Shack Rabbit," with its obvious connotation, was an A-2 jacket owned by Lt. L.F. Taylor. Unfortunately, not much else is known about the owner other than that he served in the 381st Bomber Group, 533 Bomb Squadron. They flew B-17s out of Ridgewell, England, and by the style of the U.S. insignia on the back, he served early on in 1943. He flew some 25 missions, which was the requirement in the early phase of the air campaign against Germany.

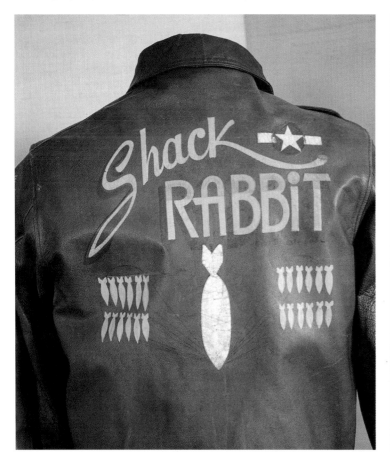

"Shack Rabbit" is a beautifully designed and scripted jacket, with national insignia to the right. Below "Shack Rabbit" are 24 small bombs and one very large one indicating the 25th and final mission. In the early phase of the Eighth Air Force air campaign, this was an extraordinary achievement, as long penetration missions were without fighter support. No wonder the large bomb was used as an exclamation point.

The 533rd Bomber Squadron insignia, beautifully executed in chenille of a skull sporting a Trojan helmet crossed by a bomb.

New Photo Gallery (Courtesy of Leighton Longhi)

Millard Wilkerson

Sgt Millard Wilkerson was a flight engineer/gunner who served in the 715th Squadron of the 448th Bomb Group, Eighth Air Force. He went to England in January 1944, completed 35 missions, and returned to the states in the autumn of that year. He was attached to several Air Force units in the states, but in April 1945, with flight engineers in short supply, he was reassigned to Air Sea Rescue service. His assignment was to the 6th Emergency Air Service Rescue squadron, ironically an Eighth Air Force unit now in the Pacific searching for downed B-29 crews. Wilkerson, a talented artist, painted all of his jackets!

A rare vignette of artist's inspiration and subject matter. Wilkerson in his bunk contemplating his next work.

Painting on Wilkerson's A-2 of his B-24 "Elie" in the 448th B.G.

Close-up photo of Wilkerson's A-2 jacket with his name tag and patch from his gunners' school in Texas.

Back of Wilkerson's M-41 field jacket with a B-17 flying over Mt. Fuji and a gate to a Shinto shrine.

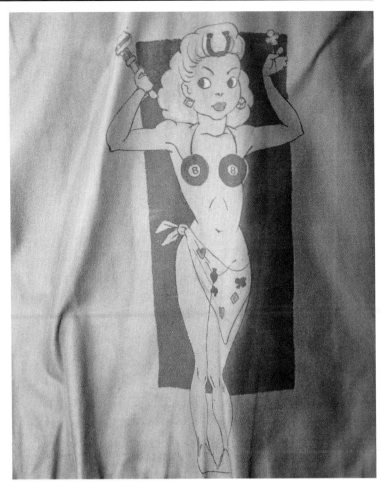

Back of Wilkerson's jacket of "Lady Luck," from Wilkerson's days with the 448th B.G.

Photo of another leather jacket of Wilkerson's, a private purchase, "Hells Angels."

Color photo front and back of another of Wilkerson's Air Sea Rescue jacket with squadron insignia and a B-17 superimposed over the Japanese islands.

New Photo Gallery (Courtesy of Leighton Longhi)

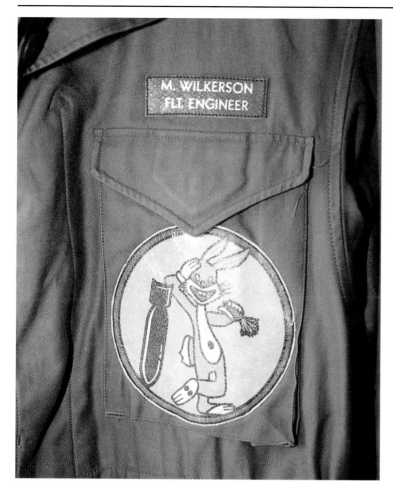

Yet another drawing on one of Wilkerson's M-41 jackets, in black with a posturing beauty.

Detail of the jacket with young girl's dress blown up into the air by flak salvoes. This was an often used pin-up model and one will find her on many other bombers with different titles.

"Shade Ruff"

This is a close-up of an A-2 jacket worn by 1st Lt. James C. Via, a B-17 pilot who flew "Shade Ruff" for the 401st Bomber Group, 614th Bomb Squadron, of the Eighth Air Force based at Kettering Midlands, England. The 401st had some of the most beautifully executed jackets in the Eighth, and the group year book has a number of pages dedicated to just A-2 jackets.

"Ugly Duckling"

Sergeant Robert M. Radlein and his jacket are pictured on pg. 181 of this book. The picture was taken outside of a B-26 reunion. After Mr. Radlein passed along his jacket he related that he served aboard "Ugly Duckling," which can be observed in the painting on the back of his jacket. The saga of the "Ugly Duckling" is one of the most interesting of the 9th AF and has been recounted in great detail in many books, in that it was attacked by JV 44, Adolph Galland's group, and made up almost entirely by experts flying Me 262 jets.

Strike photo of the Memmingen mission, apparently never published before.

Photo of Radlein's plane "Ugly Duckling" taken in late '44 or early '45 at Athies-sur-Laon, France.

New Photo Gallery (Courtesy of Leighton Longhi)

Photo of Radlein at his position in the April 16, 1945, mission to Kempten, four days before Memmingen.

Right: Radlein, taken during the Battle of the Bulge, it shows the severe weather conditions that confronted all personnel at that time. Radlein related that he and his crew wore the English quick release harness (as shown here), but quickly returned to the American harness after their bombardier plunged to his death after the faulty use of one.

"Booger Red II"

Captain Charles B. Skipper III, pilot of "Booger Red II," served in the Army Air Corps from 1942-1945, flying 67 missions in B-26s. He was attached to the 559th Bomber Squadron, 387th Bomber Group of the Ninth Air Force. After the war he flew less hazardous routes for Pan American Grace Airways. Skipper was awarded the DFC and the Air Medal with 11 oakleaf clusters, as well as a Presidential Unit Citation and ETO Medal with four battle stars. His last mission was on D-Day.

Close-up of Skipper standing on top of "Booger Red II" (first figure on left).

Skipper related that the origin of the name "Booger Red" came from an old North Carolina folk story. The jacket is one of the more impressive A-2s in existence, and is typical of B-26 unit painted jackets, which always seem so individualistic in style.

Left: Close-up of "Booger Red II" nose art, which is being published for the first time. Previously it has always been reproduced by an artist (somewhat inaccurately).

"Nyes Annihilatiors"

Sergeant Harry "Lee Gray" was a tail gunner who flew 78 missions and was assigned to the 449th Bomber Squadron, 322nd Bomb Group, Ninth Air Force. After flying 63 missions he returned to the States in May of 1944. Apparently he missed combat flying and returned to the group in August of 1944 to fly another 15 missions. Gray flew some of the most famous aircraft in the ETO: "Hank Yanks" (29); "Murder Inc." (14); and "Sit Git" (1). He also flew two missions on "Flak Bait," now in the Smithsonian and holder of the ETO's most combat missions at over 200. Gray received the DFC and Air Medal with two silver and three oak leaf clusters, 4 battle stars to his ETO ribbon, and the Distinguished Unit Badge.

Close-up showing a 322nd B-26 with a highly impressive 78 missions. The maker of this A-2 is the rarely seen David Doniger.

Lee Gray at work. Being a rear gunner was lonely, but it was perhaps the most important defensive position on an aircraft.

Lee Gray (first on right) and the crew of "Hanks Yanks"

"Miss Betty Jean"

Lt. Jack R. Fehrenbach was a P-38 pilot and squadron leader of the 94th Squadron of the 1st Fighter Group, the famous Hat in the Ring Squadron from WWI. They were attached to the 15th Air Force in Italy.

Back of Gray's jacket, "Nyes Annihilatiors" was the *nom de guerre of t*he group, named for their commander, Col. Glen C. Nye. Gray related that Ted Simonaitis, who painted all the nose art in the group, painted his jacket.

New Photo Gallery (Courtesy of Leighton Longhi)

A great shot of Fehrenbach wearing his A-2 jacket standing in front of "Miss Betty Jean," Foggia, Italy, 1944. Inside of Fehrenbach's jacket a note was found "To anyone who owns this jacket after March 6, 1996. Please love and respect World War II veterans—they kept us free. The jacket was worn by Jack Richard Fehrenbach—my dad, my hero. Betty Jean is my mom."

This spectacular jacket shows a P-38 gaining altitude in the afternoon sunlight, catching the upper surface of the aircraft. The name "Miss Betty Jean," Fehrenbach's future wife, appears below the name of the aircraft. Because of its professional style, the painting was most likely done by a local Italian artist, as were most of the 15th Air Force jackets. The Italians set up businesses catering to U.S. airmen and executed paintings for jackets in addition to personalized scarves, whistles, pillow covers, and other items.

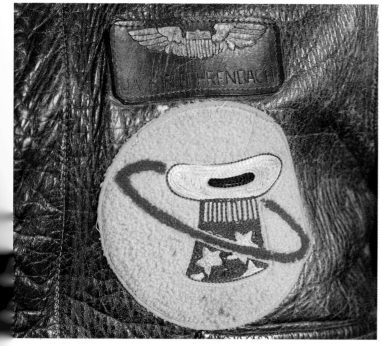

Close-up showing a 322nd B-26 with a highly impressive 78 missions. The maker of this A-2 is the rarely seen David Doniger.

Front view showing Fehrenbach's A-2 with name tag incorporating silver washed impressed pilot wing. Below is a beautiful 94th chenille patch. The right side of the jacket shows the American flag, and on the left the 15th Air Force patch. American shoulder flags were quite common on 12th and 15th Air Force jackets, and perhaps served as an I.D. if they had to bail out in the hostile environments encountered in Eastern Europe.

AMERICAN FLIGHT JACKETS, AIRMEN & AIRCRAFT

Bill Cavanaugh

Bill Cavanaugh was a Bombardier with the 460th Bomber Group, 761st Bomber Squadron, 15th Air Force. He received 2 DFCs and was the lead Bombardier on a large scale mission that enabled the Allies to break the German lines in Northern Italy, 1945. The 460th took part in numerous missions against targets in Italy, France, and Germany, as well as all of Eastern Europe.

Italian made applied painting with a B-24 silhouetted by a waving American flag in a flak-filled sky. Bill Cavanaugh related that such paintings were commonly sold by Italian vendors who set up their business close to U.S. airfields.

Shot of 460th Bomber Group B-24 in action.

Front of Cavanaugh's jacket, showing an Italian made hand tooled 460th "Black Panther" leather patch.

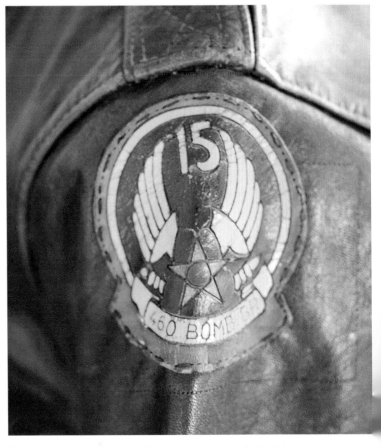

Close up of the 360th Bomb Group patch on the left shoulder.

New Photo Gallery (Courtesy of Leighton Longhi)

A cartoon-like painting of a P-40 pilot chasing an escaping axis aircraft (not seen) with a 45 pistol. He is riding in a P-40 camouflaged for North Africa. Judging by the national insignia, this pilot saw service in 1942/43, making it the earliest painted jacket recorded in these books.

Detail of the 65th Fighter Squadron with a fighting cock and the call letters of Lt R R Barnard's plane, "46." Unfortunately, at present nothing is known about Barnard's career with the 57th.

"Hold it Pal"

"Hold it Pal" was a jacket worn by a pilot of the 57th Fighter Group, 65th Fighter Squadron, who flew P-40s in North Africa. At that time the group was assigned to the 12th Air Force.

A pair of desert boots, sporting the insignia of the various squadrons of the 79th Fighter Group, also a part of the "Desert Air Force." These boots were most likely done at the same time as Barnard's jacket.

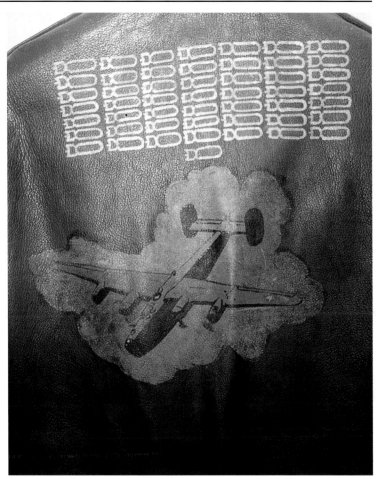

A B-24 is painted in darkened colors flying throgh a moonlit sky. Above the B-24 one can see 50 bombs, an astonishing number of missions for heavy bomber crewmen, even though most of the missions were dedicated to the supply of the resistance and partisans. A small burnt spot on the right sleeve, where hot flak had embedded itself, shows that such missions could be extremely dangerous.

Tech Support Clarence E. Craver was a radio gunner who took part in "Operation Carpet Bagger." He flew combat missions into France, Denmark, Italy, and Yugoslavia from July 30, 1944, to April 24, 1945. He was assigned to the 15th Provisional Special Group, 859th Bomber Squadron. Originally, he was assigned to the 492nd Bomber Group, then to the 15th Provisional Group, and later to the 261st Special Group. Craver and his unit were top secret. They were told to remove their insignia, which is evidenced on the front of the jacket, and were instructed not to speak about their operations even after the war. Craver related that they were briefed only on what they had to know, and before night missions people were mysteriously driven up and loaded on their aircraft. He also mentioned that they made several landings in France to drop off and pick up passengers and supplies—no mean accomplishment for a B-24.

A picture of Craver, second from left, in front of his B-24 "Mary Jane Ace of Spades."

New Photo Gallery (Courtesy of Leighton Longhi)

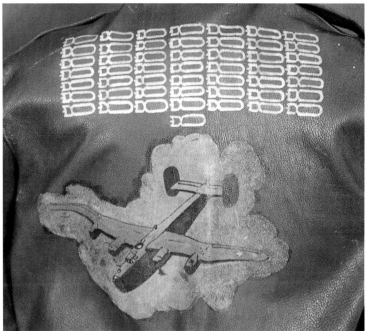

Close up of Craver's jacket.

Left: Another photo of Craver.

"The Old New Men"

T. Sergeant Harold Scherer was an aerial gunner on a B-24 named "Old New Men" and assigned to the 756th Bomber Group, 747th Bomber Squadron at Cerignoca, Italy, and later at Stornara, 15th Air Force.

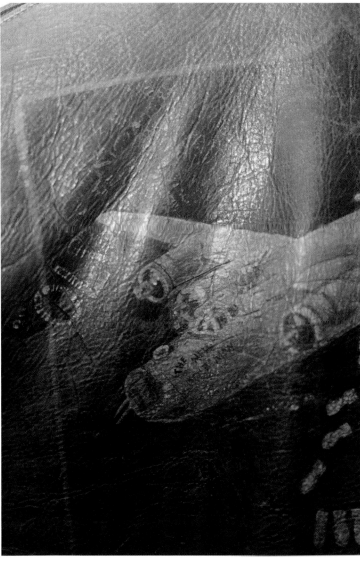

Hal Scherer's A-2 was thought to be painted in early 1944 by an Italian artist in the area of their base at Stornara. A B-24 is painted in flight on its bomb run. It is typical of those done by professional Italian artists.

Close-up of the nose art showing the "Old New Men."

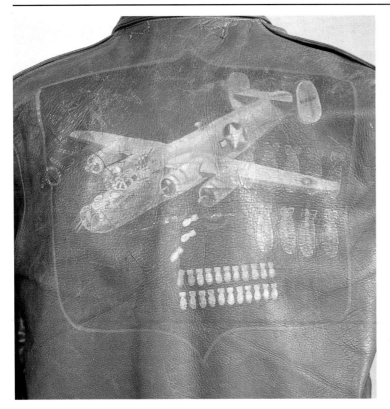

Another view of Scherer's jacket art.

Scherer at his position as a waist gunner on their B-24.

Harold Scherer at the 756th Bomber Group on a visit from Humphrey Bogart (Scherer pictured to the right of Bogart).

New Photo Gallery (Courtesy of Leighton Longhi)

Lt. (later Captain) William Otto was a pilot instructor with the South East Training Command, stationed at various fields throughout the South. Otto's brother flew P-38s in the Pacific with Major Bong's group (America's leading ace). Otto finished the war at Tinker Field, Oklahoma City, Oklahoma. He trained pilots in the venerable AT-6, and without such dedicated instructors, America could have never launched the massive air fleets that were sent overseas that turned the tide of victory.

Photos of Otto's goat skin A-2 with name tag and South East Training Command patch.

Otto (third from right) with a group of fellow instructors.

A group of AT-6s form Otto's training flight.

Also from the publisher

Gooney Birds and Ferry Tales
The 27th Air Transport Group in World War II

Jon A. Maguire and the Men of the 27th ATG

The 27th Air Transport Group, part of the 302nd Transport Wing, supported the 8th and 9th Air Forces in World War II with ferry and transport services. Though their role was extremely vital to the success of the air and ground wars in Europe, their story has remained largely untold ñ until now. Flying primarily C-47s the 27th ATG performed a broad assortment of duties including resupply of frontline units, medical evacuation, transportation of VIPs and many others. The 27th supported Pattonís drive across Europe by hauling gasoline ñ an extremely hazardous undertaking ñ to the front lines. On Christmas Eve 1944 the Group flew a special, all-out mission to transport reinforcement troops from Marseille to the Battle of the Bulge front ñ this operation involved over 100 aircraft with the 302nd Wing receiving a commendation from Gen. Spaatz for their efforts. Elements of the 27th also participated in a secret mission to Sweden to support Norwegian underground forces ñ both American and German forces used the same Swedish airfield! Many first-person anecdotes, over 600 photographs, and a reprint of the ìofficialî 302nd Wing unit history make this volume, by Group historian Jon Maguire, a fitting, and long-overdue tribute to the men and women of the 27th ATG.

Size: 8 1/2" x 11": over 600 b/w photographs, 12 color aircraft profiles
352 pages: hardcover
ISBN: 0-7643-0592-1　　　　　　　　　　　　　　　　　　　　　　　$59.95

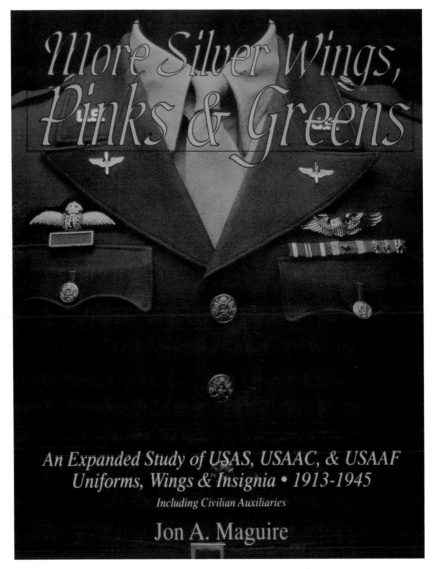

More Silver Wings, Pinks & Greens
An Expanded Study of USAS, USAAC, & USAAF Uniforms, Wings & Insignia - 1913-1945 Including Civilian Auxiliaries

Jon A. Maguire

Jon Maguire's book Silver Wings, Pinks & Greens was a tremendous success and set a new standard for American uniform and insignia references. Following its publication there still remained among collectors and historians a desire for more information. There was also an outpouring from many serious collectors who made available a wealth of items-the result, More Silver Wings, Pinks & Greens. This book of all new material greatly expands on the wing qualification bagdes, uniforms, and patches presented in the first book. Additionally, this work covers totally new areas including Civil Air Patrol, W.A.S.P.s, Air Transport Command, Factory Techincal Representatives, and "Yanks" in the RAF an RCAF. Other new areas presented are uniforms and insignia of the First World War era, and the "Golden Age" of the 1920s-1930s. There is also a large section on Aviation Cadets and civilian contract flying schools and instructors. The book is presented in the detailed and thorough style typical of Jon Maguire's work. Original items are shown in over 1000 color photographs, as well as numerous unpublished period photographs showing the items as they were worn.
Size: 8 1/2" x 11": over 1,100 color and b/w photographs
350 pages: hard cover
ISBN: 0-7643-0091-1 $79.95